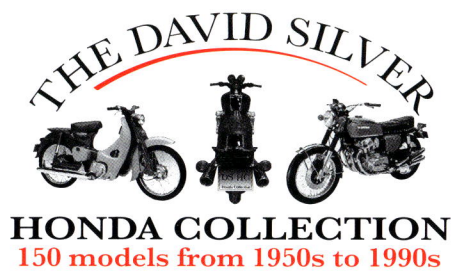

HONDA COLLECTION
150 models from 1950s to 1990s

YOUR GUIDE TO EARLY HONDA MOTORCYCLES AND
THE DAVID SILVER HONDA COLLECTION

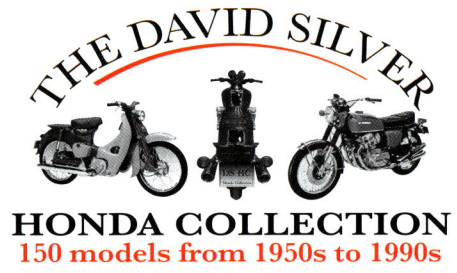

HONDA COLLECTION
150 models from 1950s to 1990s

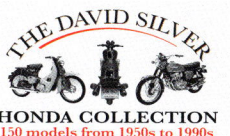

Author: Roger Etcell
Creative design and artwork by: Susan Whatling
Printed and bound by: Leiston Press Ltd, Station Road, Leiston, Suffolk IP16 4JD
Copyright: David Silver Spares Limited
First published: February 2017

ISBN number: 978-1-911311-09-6

Published by: David Silver Spares Limited
Unit 14, Masterlord Industrial Estate
Station Road, Leiston
Suffolk IP16 4JD
United Kingdom

+44 (0) 1728 833 020

sales@davidsilverspares.co.uk
www.davidsilverspares.co.uk
www.davidsilverhondacollection.co.uk

All rights reserved.
No part of this publication may be reproduced or transmitted in any form or by any means without prior approval in writing from David Silver Spares Limited.

FOREWORD

Motorcycle museums can be a bit dull, really. Packed with stuff that's either so old or so exotic that it's hard to sense any real relevance to your own life as a bike enthusiast.

The David Silver Honda Collection is a bit different. Yes, there are a few rare and unobtainable things in here, but on the whole it's a story of popular bikes owned by normal bikers. It just so happens that some of them – the Super Cub, the CB750 Four, the Fireblade – are also deeply significant to the story of the motorcycle, but that's because they're Honda, and Soichiro Honda was one of those rare visionary engineers who was able to change the world with a machine.

Despite being located in the turnip-growing region of Suffolk, rather than a more established industrial centre, the David Silver Honda Collection is one of the largest concentrations of Honda road bikes in the world. This book is your guide to a walk-around of over 150 bikes that should occupy you for a very relaxing and evocative half a day or so, and during which you will probably come across a bike or two that you once owned.

More to the point, you might come across a perfect and 'historically correct' example of one you're trying to restore, which makes this collection a sort of primary resource for Honda fans. I've restored a few myself, and the missing bits have always come from David Silver Spares. That's another great thing about this museum; you can work out what you need for your classic Honda, and then buy the bits from the handy parts department next door.

I hope you enjoy the collection, this book, and the opportunity to meet some like-minded fans of the work of the great Soichiro. After all, you meet the nicest people in a Honda museum.

James May

▶ James enjoys the CB400F during the 29th June 2016 official opening of the museum. Joined by David Silver and three-times World Champion for Honda, Freddie Spencer.

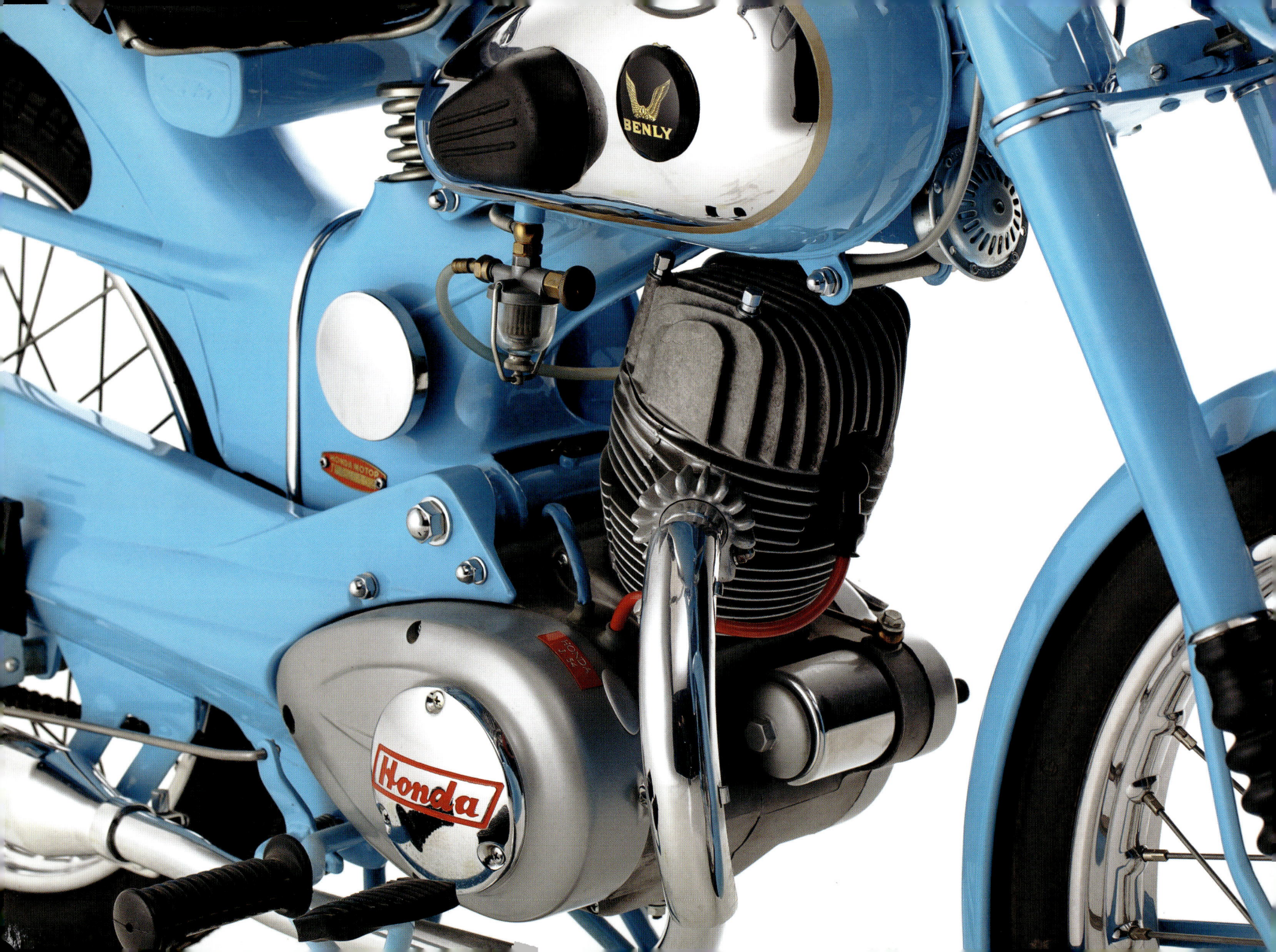

CONTENTS

1.	Introduction	Page 8
2.	About David Silver & David Silver Spares Ltd	10
3	About the Author	12
4.	About Soichiro Honda	14
5.	Honda Motorcycles, in the beginning	16
6.	The 1940s	17
7.	The 1950s	24
8.	The 1960s	44
9.	The 1970s	64
10.	The 1980s	84
11.	The 1990s	104
12.	Today and Tomorrow	108
13.	General Specifications of the Models on Display	110
14.	Visiting us…	114

◄ The 1954 Benly J-Type. One of the most intriguing early Honda models on display in our museum.

INTRODUCTION

In 2013, I acquired a large assortment of classic Honda bikes from a closed private collection museum in the USA. I was always keen to add these to my own collection of Hondas, which have been assembled over many years, to create a permanent and public exhibition at our Leiston, Suffolk base for enthusiasts to enjoy.

By early 2014, I had made the commitment, The David Silver Honda Collection was to be established. Work started on a new two-storey building, next to our parts business, to be ready for a grand opening in the summer of 2016. Established local builder (and Honda rider!) Charlie Ridgeon, project managed and built the building and local creative designer, Susan Whatling crafted our corporate identity.

Roger Etcell, former Head of Honda UK motorcycles, past Honda dealer and now a Honda historian, joined me to project manage the museum creating the display layout and all the historical and model information. He has also now produced this Guide Book, together with Susan's creative design, to complement the museum display.

Formally opened by TV presenter James May on the 29th June 2016 and featuring over 150 classic Honda motorcycles from 5 decades, The David Silver Honda Collection is my personal collection that is presented in logical year order of introduction. As you stroll through both the lower and upper floors, the history timeline of Honda is depicted on a wall frieze enabling you to gain an understanding of not only what the Honda company had in mind each year, since their very beginning, but also appreciate the speed of their development in becoming the world's largest motorcycle producer.

You will see some interesting Honda models in the display including some rare early 1950s models, hardly ever seen outside of Japan. Also, there are some US models not sold in the UK and a few others, which represent the first of their type off the production line.

The Super Cub C100 and Dream CB750 are shown together on our feature display, as they are very significant models of the entire Honda range. The early success of the Super Cub C100, which continues today with over 80 million units sold, enabled the company to resource their 1960s racing program, which in turn gave them the successful technology for models such as the Dream CB750, which in itself was recognised as the founder of the term Superbike! In other words without such success of the C100 we may never have seen the CB750!

The earliest Honda on display is a 1951 model Dream E-type produced just a few years after the company was founded. Our latest model is a 1992 Fireblade where we currently finish the collection. This represents not only the period of probably the most exciting model development, but also the period during which, the founder and president, Soichiro Honda, started and finished. He passed away in August 1991 at the grand age of 85!

A good number of the bikes in our collection, mostly the US models, came from Bob Logue Motorsports in

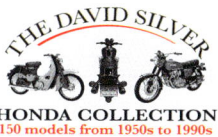

Pennsylvania. Bob started as a Honda dealer in the 1970's and each time a customer traded in a Honda he added it to his private collection. By the time I visited his store in 2013, he had amassed around 130 bikes. His mechanic of some 40 years, Myron Schroeder (in his 90's), was still restoring Honda motorcycles. It was at this time that Bob and myself agreed to a sale of around 100 of the bikes, which we imported into the UK. Bob was keen on the idea that his decades of collecting Honda were being kept as a collection, for enthusiasts to enjoy. Myron sadly passed away the following year, but his years of dedication to classic Honda live on in this collection.

Even with over 150 different models on display at the time of publication of this book, there remain a number of gaps to fill and over time the collection will be further extended as additional Honda models are sourced.

I hope you enjoy the display.

David Silver

About David Silver & David Silver Spares Ltd

David, North London born and educated in the 1960s/70s, attended Lycee Francais (a school for the children of French diplomats) and Mill Hill School before gaining a BSC in chemistry at King's College London in the early 1980s.

Post University, David worked for his father's company as a sales representative selling kitchen gadgets to wholesalers and retailers, gaining important knowledge and experience in general business practices.

In the mid to late 1980s, a Saturday job at bike dealer Sondel Sport of Highbury Corner North London, introduced him to the motorcycle business and it was here that he gained knowledge of the intricate Honda parts numbering system. He also found where all the hoards of unwanted old parts were located; such as Honda parts distributors Parks of Lewisham, Fowlers of Bristol, Tippetts in Surbiton and Lings in Norfolk to name but a few.

In 1986 and still in North London, David built the business that bears his name with the supply of new old stock (NOS) parts for Honda motorcycles purchased from these parts distributors and other franchised dealers. The very first purchase was a couple of hundred pounds worth of obsolete NOS petrol tanks and mudguards.

A few years later and a move from London to Leiston in Suffolk gained David some much needed warehouse capacity and the opportunity for the fast growing new business to fly.

Travelling the length and breadth of the UK as well as Europe and North America brought David in touch with extensive stocks of NOS Honda parts. The company soon became known for the supply of obscure new parts enabling owners of early Honda models to maintain and restore their early bikes.

Today, David Silver Spares Limited carries a mountain of parts for more than 1,000 different models at both his Suffolk business and, as from 2012, at his similar USA business in Pennsylvania.

The company's extensive inventory includes one of the most comprehensive range in Europe for Honda models of the 1960s and all the way through to the 2000s.

David Silver Spares Ltd is not a franchised Honda dealer for new bikes but a company that has focused on supplying new parts to help owners keep their classic Honda on the road and in prime condition. The most popular models for parts requests are the 1970s SOHC fours such as the CB400F and CB750, the CX500 and CB250/400N Super Dream twins as well as the GL Gold Wings.

◀ David with one of his favourite Honda models, a 1970s CB400F.

Growing in demand today are parts for the later 1980s/90s models such as the in-line four CBR range, the V4 VFR models and ST1100 Pan European. Also, Japanese home market models such as the CBR250/400, VFR400, RVF400 and CBX750 Bol'dor and the list goes on to many other European, Canadian and USA imported Honda models.

If you are looking for a spare part for your classic Honda motorcycle, the chances are David Silver Spares either has it or can obtain it for you.

In addition, David has restored, built and sold many CB400Fs, something he became famous for in recent years. He regularly has many classic Honda bikes for sale as working examples or as project bikes for enthusiasts to restore at their leisure.

The museum collection is the latest addition to David's business, adding yet another dimension to his Honda portfolio. His personal favorites in the collection are the Benly J-type, SS50 and of course the CB400F.

With Honda parts businesses in the UK and the USA and the museum collection now open to the public, who knows what's next for David Silver, watch this space as they say!

▼ A small section of the parts warehouse at David Silver Spares.

About the Author

Roger Etcell has had Honda motorcycles in his blood for most of his life! From the start of his biking interest in his mid teens, he has now ridden Honda bikes for over 45 years.

Having gained initial industry experience working for the past Royal Enfield famed Gander & Gray dealership in East London as well as the Essex based Johns of Romford dealership, both of which held Honda franchises; Roger joined Honda UK at Power Road, Chiswick in the mid 1970s.

Based initially in the Service Department, he soon moved to Sales and Marketing working for the formidable Sales Director, Eric Sulley. These were the heady days of Honda sales with new models arriving every month. Eric was a dominant figure in the Industry, which Honda led with over 45% market share and over 600 dealers across the UK.

Roger's role was varied; visiting dealers, establishing ad campaigns, creating new sales brochures, working on sales projections and liaising with Japanese colleagues. Also producing the annual dealer seminars as well as managing the bike press fleet and attending new model planning meetings in Japan which included test riding new prototype models.

Eric Sulley was the king of motorcycle sales in the 1970s and early 80s and Roger's time with the ex-military, smartly dressed, flamboyant Sales Director was a time he will never forget. For readers who may recall Eric, they will know what Roger means!

In 1979, it was decided to go on TV with the all-new Honda Express moped. Twiggy was chosen as the ideal celebrity for the commercial but on arriving to the film set she didn't realise she was required to ride the bike. Not having ridden before, it was Roger's task to teach her! As Roger says; 'I could have done it in 20 minutes but I did make it last a few hours!'

Roger's role in new product development took him to many meetings in Japan and across Europe to review new prototypes and test them on the company's banked test circuit in Japan or on the unrestricted autobahns of Germany. As he says; 'a very privileged and satisfying role, especially when you see the final product come to fruition'.

One of his memorable product developments was the creation of the VFR750F in late 1985. By the end of 1984 the relatively new VF (V4) range had troubled PR with its well-publicised cam-drive problems. R&D wanted to cancel the V4 range

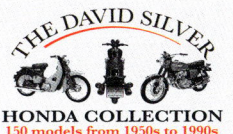

and revert to the traditional in-line four. Roger believed this was wrong and, along with his German counterpart, spent many a late hour in a hotel, meeting with senior R&D members convincing them to persevere with the V4 and revert with an all-new model, 110% better. The rest of Europe and the US markets were not so keen but the UK and German representatives won the day. R&D went back to the drawing board and the 1986 VFR750F was later launched. The rest as they say is history!

On reflection Roger sees those days at Honda as very special times, the best times, the times when business was done with enjoyment and satisfaction and honored on a handshake.

By the mid 1980s, Eric had left the company and his successor Gerald Davison had also left which handed Roger the plum job of Head of Motorcycle Sales. He enjoyed this but after a couple of years he could see dramatic changes to business practice being implemented from Japan.

Realising it was time for a change, Roger left the company to start his own Honda dealership in North London. He took over an existing Honda franchised dealership and set about changing its ageing appearance and re-named the business 'The Bike Studio'. It was one of the first modern style showrooms with carpeted floors, coffee lounge and courtesy bikes for all customers.

He recalls selling many Honda bikes, including a number to celebrities such as Andrew Ridgley of Wham and Pink Floyd drummer Nick Mason. One particular sale was quite special. Roger had been allocated one of only 32 NRs when they came to the UK in 1992. A customer, coming from Belfast, arrived with £38,000.00 cash (the then list price), in two Tesco shopping bags, ready to take the cherished bike back home!

By 1994, Roger had sold his business and took up the offer of Sales & Marketing Director for Silverstone Circuits Ltd. He worked with a small board of directors to repair the declining business to a profitable one within 5-years. The burgeoning business was then bought by an American corporation who made the senior management, including Roger, redundant. They went on to lose £5m in the same number of years!

From 2001 Roger has provided automotive marketing consultancy and event promotion services. This included managing the annual London to Brighton Veteran Car Run and creating the Brighton to London Future Car Challenge and Regent Street Motor Show through his company Motion Works Management Ltd. He also diversified to provide transport logistics consultancy to the 2015 Rugby World Cup. He then returned to his roots to project manage the layout and historical information for David's museum as well as writing this book. Roger also continues to enjoy his passion of restoring classic Honda motorcycles as well as riding his current road bike, a Honda VFR800 of course!

Soichiro Honda - founder of the Honda Motor Company, a world-renowned pioneer in automotive engineering and mass-production. He was a man with enormous vision, passion, ambition, pragmatism and tenacity.

From very humble beginnings in the late 1940s, it is today hard to miss a Honda product in our daily lives. Be it a moped, motorcycle, car, lawnmower, generator, marine engine, agricultural machine or even a Honda robot or Honda passenger jet aircraft!

Born in 1906 in Tenryu, a small village under Mount Fuji near Hamamatsu City in Japan, Soichiro spent his early childhood helping his father, Gihei, a blacksmith, with his bicycle repair business. His mother, Mika, was a weaver. His grandfather would often take him to see a machine working in a local mill and in his early teens Soichiro borrowed one of his father's bicycles to visit a local airfield to see a very rare demonstration of an airplane. These experiences cemented Soichiro's love for machinery and invention.

With a poor school record, he was not interested in traditional education, Soichiro left school and home at the age of 15 and

◄ Soichiro Honda (1906 – 1991).

About Soichiro Honda

headed to Tokyo to look for work. He found employment as a car mechanic at an established garage. He stayed for six years, surviving the Great Kanto earthquake of 1923 that claimed over 140,000 lives, and in his early 20s returned home to start his own auto repair business.

Soichiro repaired the cars of local residents and gained a reputation for fixing problems that others could not solve. He started car racing and competed with a turbocharged Ford in the "1st Japan Automobile Race" at Tamagawa Speedway in 1936 but crashed with some serious injuries. After that, he quit racing!

In 1937, after a period of recuperation, Honda founded a new company, Tōkai Seiki to produce piston rings. The products initially had a very high failure rate so Soichiro went back to school as a mature student to learn more about metals. In 1944, during World War II, a US B-29 bomber attack destroyed his Yamashita factory and a year later, his Iwata factory collapsed in the Mikawa earthquake. Post war Soichiro sold the salvageable remains of the company to Toyota and took time out to enjoy life for a while and to reflect on what to do next.

During 1946 and from the sale proceeds, Soichiro founded the Honda Technical Research Institute to design and produce tooling and develop a method for attaching a small petrol engine to a pedal-cycle. Work was primitive; an early development model used a hot-water bottle from his home as a fuel tank! He had purchased 500 units of Imperial Army surplus generator engines, used in the fields for radio communications, manufactured by Mikuni Shoko, the same company that became famous for producing carburettors. From an early success, he quickly sold out of the 500 engines and started work on his own replacement engine.

By September 1948, the company was renamed Honda Motor Company Limited. He then employed Takeo Fujisawa, a previous business associate and friend, to run the sales and business side of the company. Soichiro worked relentlessly at the factory on new models and creative production engineering, rapidly promoting the company to the worlds largest motorcycle producer in just 11 years!

By the early 1960s, the company had over 5,000 employees and was producing motorcycles at the rate of one every 40 seconds! The factory production systems were highly automated and the JIT (Just In Time) method of parts supply to the production line was already established.

Soichiro Honda remained president until his retirement in 1973, where he stayed on as director and then supreme adviser from 1983. His status was such that People magazine placed him on their '25 Most Intriguing People of the Year' list for 1980, dubbing him the 'Japanese Henry Ford.'

In retirement, Honda busied himself with work connected with the Honda Foundation and even at his advanced age, Soichiro and his wife Sachi both held private pilot's licenses. He also enjoyed skiing, hang-gliding and at the age of 77, ballooning. He was also a highly accomplished artist.

In August 1991, Soichiro Honda died of liver failure. He was posthumously appointed to the senior third rank in the order of precedence and appointed a Grand Cordon of the Order of the Rising Sun.

Honda Motorcycles, in the beginning...

It's the mid 1940s and war torn Japan was in a bad way and desperate for daily transport, both public service and personal. Roads and rail were in poor condition and public transport was in short supply and unreliable in service. Daily commuters walked for miles to get to their place of work and some cycled, as pedal cycles were the only means of personal transport. Fuel was almost non-existent other than limited supplies on the black-market!

The Empire surrendered to the Allies in 1945 and a period of occupation by the Allies followed. A new constitution was created with American involvement in 1947, officially dissolving the Empire with occupation and reconstruction continuing well into the 1950s. The country eventually reforming as the 'State of Japan.'

There were a number of motorcycle and scooter manufacturers in Japan pre WWII and many more post war but few were to survive. From the established list today, Honda started in 1946, Suzuki in 1951, Yamaha in 1955 and Kawasaki in 1961.

Across Asia, motorbikes had been primarily shipped in from the US and Europe. These were in high demand and in some cases were used as examples to copy from, for home-market production ideas.

In Europe, hundreds of pre-war motorcycle and scooter manufacturers had existed mainly in England, Germany, France and Italy. Many of these producers had switched production to military hardware during the war and only some reverted back to motorcycle assembly post-war. But the post-war future looked good for powered two-wheelers and soon demand was high for both motorcycles and scooters.

The 1940s - 1946

At the same time in North America, large capacity touring bikes from US makers Harley Davidson and Indian were in demand for primarily leisure use. This market was joined with Triumph, BSA and Norton imports from England and with BMW and others from Europe. The mass use of powered two-wheelers for daily commuting had not been conceived at this time.

Its 1946 and back in Japan, Soichiro Honda, who by now had experienced many years in the automotive industry, could see the need for personal transport. He could also see through his father's bicycle repair business, that the daily use of pedal power was taking its toll on the riders. With his interest in anything mechanical, he explored the idea of adding motorised power to pedal cycles. This was not a new idea but something that he felt could be further exploited.

Soichiro came across the opportunity to purchase 500 ex Japan Army war surplus generator engines that were used for powering radio equipment in the field. After many experiments with one sample, including using a hot-water bottle from the family home as a fuel tank, he succeeded, albeit in a fairly crude fashion, in mounting one of the engines to a cycle frame and create a method to drive the front wheel. Due to rapid tyre wear this system was soon converted to drive the rear wheel via a v-belt. Petrol was still rare and expensive, so an alternative fuel was needed.

A turpentine-based oil extracted from pine trees was used even though this caused difficult starting and an unwelcomed smelly and smoky exhaust!

From these humble beginnings, the Honda business started.

October - Soichiro Honda established his new business with the grand title of **Honda Technical Research Institute**. Work started on developing and producing machine tools as well as developing internal combustion engines for use on pedal-cycles.

The 500 war-surplus Japan Army generator engines were purchased. Soichiro and his new team developed an acceptable mechanical system to install them into used and customers own pedal cycles, providing a very welcomed method of powered transport.

Based in Hamamatsu, the company's first factory was a small 25ft x 30ft wooden shack with 12 employees working 6 days per week.

Soichiro's wife, Sachi, was given one of the early prototype powered bicycles to use and feed back comments. She found that her clothing became soiled in oil whilst riding and this problem was taken on board with improvements made. Sachi Honda effectively became the first female test rider!

Word of the engines for pedal-cycles soon got around and customers came from far and wide to purchase a complete cycle with engine installed or just the engine and fitting kit. It wasn't long before all 500 of the war surplus engines had been sold.

▲ The auxiliary engine bicycle.

▲ Early production facility.

▲ Mr Honda outside his first factory.

1947

With all the 500 ex-war surplus engines sold, Soichiro Honda was busy designing a replacement, an engine that would be a true all-Honda product. He hated the idea of simply copying; his products must have elements that are unique, different to the opposition and beneficial to the customer.

Using the floor of the workshop as a drawing board, Soichiro was seen late one night using chalk to sketch a new engine idea, a unique design that was to be known as the Chimney engine. It was produced as a working prototype but problems prevented it going any further.

The Chimney name, as it suggests, came from the fact that both the piston and the cylinder head had a long protrusion on top. There was an unusual central scavenging system, making for a most unconventional 2–stroke design. No such engine had ever been created before, but the aim with this engine was to minimize the disadvantages of the 2–stroke and to improve performance. In other words, it was supposed to reduce fuel consumption and raise power. However, the machining tolerances and materials available at the time were simply not up to the requirements of this design and the engine was subject to one problem after another.

The engineering drawings and prototypes of this engine have all disappeared, giving it something of a phantom presence in Honda history. Interestingly however, 50 years later the chimney engine came back to life when Honda engineers created a working replica from what was known, for the Honda Collection Hall museum in Japan. According to the chief engineer responsible, making the engine using today's machine working technology and materials resulted in

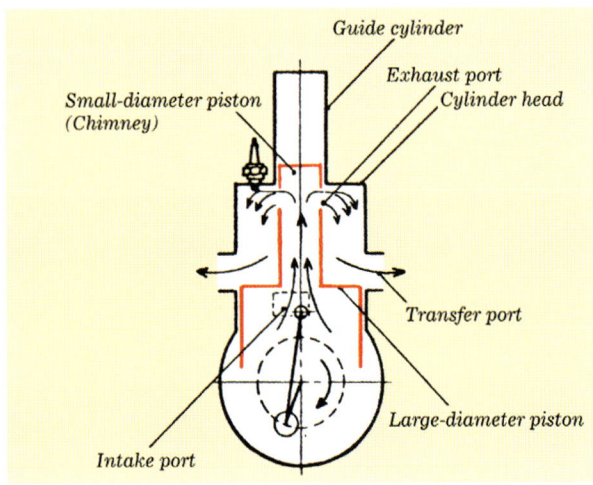

▲ The original Chimney engine design.

▲ Factory number 2, The Yamashita plant.

performance levels well beyond other 2–stroke engines of its time. Just as Soichiro had thought, it achieved low fuel consumption. Theoretically, it was a correct direction to take; it only failed because it was too far advanced for its time.

Mechanical design failure was something that would happen time and again at Honda as they continued to push engineering boundaries. However, when they stumbled they persevered, as often something good would come from it. It is the Honda way to take the experience of failure and later use it as the fuel for future success.

Finally, a more traditional looking engine design was employed but inside there were several features unique to Honda. The Chimney name stuck and the new engine fitted to a cycle, or available separately, became the company's first full product, named the A-Type.

Production of the A-Type started in November and sales immediately took off.

◂ The 50cc 2-stroke A-Type.

1948

The Honda A-Type was an immediate hit and in less than a year the product gained 66% share of the domestic market.

Busy with volume sales, Soichiro had big ideas for his new company. Not only for new products but also for the way the company would be structured and how basic engineering could be enhanced to handle volume production. His philosophy was also strong and even in these early days he would constantly remind his staff of his thoughts with short briefings such as…

▼ An original A-Type aluminium fuel tank with faint early company logo featuring a nude figure racing across the heavens.

'When you're making something, think about the person who will be spending the most time with it. And the person who spends the most time with it will be the customer, right? Next is the repairman at the place that sells the product. Next are the people in our plant. Even though you're the one who makes it, the designer spends the least time with it of all. If you put yourself in the place of the person who'll be using the product over a long time, then you won't be able to design an unfriendly product.'

The initial A-Type auxiliary bicycle engine and fuel tank is today preserved in the Honda Collection Hall. It originally had an aluminum fuel tank that was made by sand casting in two parts, upper and lower. In order to prevent fuel leaks through pinholes that often resulted from this casting technique, the tanks were given a coating of Japanese lacquer. The brand mark from Honda's founding period, which shows a nude figure racing across the heavens, is still faintly visible on the side of the preserved tank.

When the chimney engine was being reconstructed in 1996, the engineer responsible also examined its sister product, the A-Type engine, very closely.

'I was just amazed', said the engineer. 'All through the engine I saw the signs of what we call a friendly design. I noticed it when I dismantled the engine. When I remove a nut from this machine, no parts fall off anywhere.

What's going on? For example, there are locknuts on the crankshaft and speed reduction gearbox bearings. In other words, these were screws that hold on the rotating shaft, and they were designed so they wouldn't cause immediate trouble even if they somehow got loose. Either the screw wouldn't come out completely, or it was arranged so that the mechanism wouldn't break down right away if the screw did come all the way out. It was designed to hold at least until the driver realized something was not right, and the problem was noticed.

This shows a concern for safety. In those days, screws were low-precision items, and everyone accepted that no matter how well you tightened a nut, it was going to work itself loose. That must be why they thought up this design.'

The A-Type was the first product to have the Honda name emblazoned on its fuel tank, and it was very well received. Brokers would come from across Japan and wait for the machines to be finished, and buy them up. From all around the Hamamatsu area, customers would come to the plant with their bicycles and make their request: 'Please install it on this.'

Bicycle shops across Japan sold bicycles with the A-Type engine already installed and some started to try and copy the engine ending up as manufacturers. Hamamatsu became the hub for these motorised bicycles that became known as 'Pon-Pon' bikes.

The next Honda product was the B-Type, a compact cargo-carrying three-wheeler with a 90cc engine based on the A-Type. Very little is known about this model as it had some poor characteristics and was quickly cancelled.

Work then started on the next model, the C-Type. Also based on the A-Type engine, now with 96cc, increasing horsepower from 1bhp to 3bhp! It was sold as a complete engine or installed on a specially constructed motorcycle frame. At this time Honda bought cycle components from outside suppliers to complete machines.

The C-Type was Honda's first race entry and went on to win a class championship in July the following year.

In September the company was incorporated and renamed Honda Motor Company Limited and new offices were established. The Company was planning for big times ahead!

▲ The 3-wheel 2-stroke B-Type.

▲ The 96cc 2-stroke C-Type.

1949

Up until now, the company had been buying-in cycle parts to build their motorcycles from third-party companies, but Soichiro was eager to build his own complete machine.

Throughout this year, the focus was on in-house manufacturing of frames, wheels, brakes and other components as well as faster assembly production processes in order to achieve their sales targets.

In August, Soichiro's dream came true with the launch of the aptly named Dream D-Type. The new model was a 98cc 1-cylinder 2-stroke with 2-speeds producing 2.3bhp. It represented Honda's first complete own manufactured motorcycle and the first of many models to be named 'Dream'.

The engine design of the D-Type featured an unusual rear of cylinder exhaust exit and a front mounted carburettor! It was also revolutionary in that it did away with the traditional hand-operated clutch. The handlebar lever on the left looked like the clutch at first, but it was actually the front brake lever.

Operating the clutch on the D-Type was easy. Pressing down and forward on the change pedal with the front of the foot would put it in first gear. Letting go would return it to neutral, and pressing down and back on the pedal with the heel would put it in second gear. This was the first motorcycle in Japan to have a semi-automatic clutch system with a cone clutch mechanism.

At that time, all the other motorcycles manufactured in Japan used steel tube frames but the Dream D-Type used a channel frame made of pressed steel plate as good quality steel tube was short in supply. Soichiro believed he could mass produce frames quicker this way, as the design required fewer welding points. Furthermore, at a time when the market assumed that motorcycles would be painted black, Honda gave its product a beautiful maroon colour to Soichiro's personal taste.

▲ Takeo Fujisawa on the right joins Soichiro Honda.

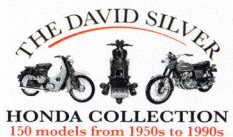

▲ The Dream D-type.

▲ Conveyor belt production line for the Dream D-Type.

The new assembly line for the D-Type used a powered belt conveyor, an original Honda design. The line was on a slope, and the assembly process moved from the higher end to the lower end to make things easier for the people working the line in terms of their posture. At a given time, a bell would ring and the line would move forward one step in the process.

The new conveyor belt mass-production process became the subject of an article in a 1956 issue of a Japanese science magazine. The article said: 'There is a company that has achieved a production increase unthinkable in our time, producing 876 units in fiscal 1950 out of the national production total of 3,439, and then leaping to 700 units a month in fiscal 1951.'

The Dream D-Type was a revolutionary motorcycle aimed at eliminating the manual clutch operations that required rider familiarity. It constituted a bold challenge to the accepted notion of the motorcycle at that time. With its bold colour it stood out on the road, and the Honda name took on tremendous appeal. Sales were very good from the start.

At this time Honda was introduced to Takeo Fujisawa, a professional businessman who had heard of Soichiro from his piston ring company days. Apparently the two men established an understanding of each other almost immediately. Their personalities were completely different and they were skilled in two quite distinct areas of business, but both men were in complete agreement as to why they got on so well: 'He's got what I haven't got.' said Soichiro. Honda was 42 and Fujisawa was 38, in today's terms they would still be considered quite young, but both already had a wealth of experience, were possessed of great powers of intuition and insight, and were excellent judges of personality.

In October, Fujisawa joined Honda Motor Company as Managing Director. Within just one month, despite an ongoing economic downturn, the company carried out its first capital increase, doubling its capitalisation to 2 million Yen. A quarter of the new money was put up by Fujisawa.

Takeo Fujisawa was to become an important figure alongside Soichiro within the Honda Motor Company and in later years became Supreme Advisor. Almost immediately on joining the Company he was left to solely establish the sales and business strategy allowing Soichiro and Kiyoshi Kawashima, who had joined in 1947, to focus on design, engineering and production.

The 1950s - 1950

No new models this year, as it was time to expand the business infrastructure.

Fujisawa had quickly set to work securing funding for a new Tokyo branch office, which would be his centre of operations, and a new Tokyo plant where production would begin later in the year. All despite the fact that the Japanese economy was again poor and the market for motorcycles was decreasing.

Honda's inventory was growing but working capital was shrinking. Payments to suppliers were in arrears and employees were being paid in instalments. Sales of the new Dream D-Type, which had sold well the previous year, were now in decline due to the recession. Talk within the factory was that the company would go bust any day now! But the more senior members believed they could help turn the company around.

In June, the Korean War broke out and Honda received a big order for auxiliary engines so the immediate crisis was over. This was a perfect example of good early business practice with the company having branched out to produce and market other products such as auxiliary engines, giving them potential security whenever motorcycle sales were in decline.

The factory was busy again and Soichiro had noted a number of employees who had worked hard to help save the company during these difficult times. He gave them share options in the company and promotion, all of which boosted moral amongst the workers. It had been an extremely difficult year for the company but both Soichiro and Takeo Fujisawa could see a brighter future and had continued to develop the business regardless.

Thanks to the surge in demand, caused by the Korean War, the Japanese economy got back on its feet and Honda had time to start thinking about its own recovery. The new Tokyo office was in operation and the new Tokyo factory was now manufacturing frames and body parts. There was a new final assembly line allowing the Dream D-Type to roll again. At this time the engines for the D-Type continued to be produced at the Hamamatsu plant and sent to the Tokyo plant for final assembly.

This difficult year was another fine example of the resilience of Soichiro Honda and Takeo. Many companies collapsed during this time of deep recession and many believed Honda would go the same way. However, Fujisawa's business foresight and strength plus Honda's determination, saw them through. In many ways this was a good year, as the pair survived an immense test and others in the company witnessed an impressive way of dealing with business downturn, management crises and survival. All of which would provide strength for business issues that would come their way in the future.

In November, Soichiro himself moved to Tokyo, returning to the capital after an absence of over twenty years. The time had come to stop being merely a provincial firm and move to the capital, a whole new world that was free of old-fashioned restraints.

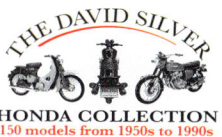

The D-Type in its bright colour finish showing its unusual rear of cylinder exit exhaust and front carburettor. Also note the front brake lever on the left, not clutch!

The Dream D-Type as on display today in the Honda Collection Hall, Japan.

25

1951

With sales of the Dream D-Type now going well, Honda and his team started work on the next new product. Soichiro wished for a 4-stroke engine to reduce noise, smoke and vibration.

In March, Kiyoshi Kawashima was summoned to the Tokyo plant from Hamamatsu, where he stayed for several months to design, from scratch, a new 4-stroke engine. By May he had the plans for the new engine, which Soichiro found interesting, as it was a radical departure from their current 2-stroke engine and he gave his approval for a prototype to be made.

In July, a prototype of the company's first 4-stroke motorcycle, the Dream E-Type with a 146cc OHV single cylinder engine and 2-speed transmission, underwent trials over Hakone mountain pass. A regular tough testing route, the bike recorded a pleasing 70kph (47mph) average speed, even in poor weather conditions! Kawashima was both the engine designer and test rider.

In October the Dream E-type went into production and almost instantly the sales demand required production output of 130 units per day, a record in Japanese motorcycle production history.

Compared to 1950, the Japanese economy and the Honda motorcycle business was now improving and the company's previous investment and development efforts were paying off.

By late 1951, the Japanese motorcycle industry had become more competitive and the market started to show preference for 4-stroke rather than 2-stroke bikes. Later, Honda became known as '4-stroke Honda.' Many other makes of 4-stroke engines were fitted with side-valves for reasons of economy and ease of manufacture, while Honda opted for the overhead valve (OHV) system. As a result Honda bikes were much more powerful than other Japanese machines.

Soichiro had really wanted to make good quality 4-stroke bikes from the very beginning. In 1940s Japan, people lacked understanding about 2-stroke engines. Since they burned oil he had only tolerated them as a stopgap, at a time when he had little money and an inadequate facility.

Other than the following years; Cub F-Type engine, which was 2-stroke for various reasons, Honda only made 4-stroke bikes for the following two decades.

The Dream E-Type was the first bike Soichiro really enjoyed making. The frame, like that of the D-Type, was of channel-frame construction. Because there had been so much trouble with the failure of the wet-cone clutch on the D-Type, the E-Type was fitted with a dry-type multiple disc clutch. The clutch control was also changed to the more conventional left-hand lever system.

Compared to the D-Type, which had seen maximum production at 160 units per month, the E-Type was regularly being produced at 500 units per month. This increased to 2,000 per month a year later when it was fitted with a third gear. Three years later, monthly production reached nearly 3,000 units.

Now that Honda had overcome the critical problems of its early years, the company would start to expand thanks to the success of the E-Type, and seize this opportunity for rapid future development.

In September, Honda published its first company magazine, 'Honda Monthly'. In December, Soichiro Honda used the pages of this publication to explain the basic corporate concept of the 'Three Joys' for

◀ The Dream E-Type, the company's first 4-stroke product.
This photo shows the actual Dream E-Type as on display in
The David Silver Honda Collection, a 1953 example.

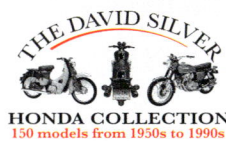

the first time. The following is a direct transcript…

'I am presenting "The Three Joys" as the motto for our company. These are, namely, the joy of producing, the joy of selling, and the joy of buying.

The first of these, the joy of producing, is a joy known only to the engineer. Just as the Creator used an abundant will to create in making all the things that exist in the natural universe, so the engineer uses his own ideas to create products and contribute to society. This is a happiness that can hardly be compared to anything else. Furthermore, when that product is of superior quality so that society welcomes it, the engineers' joy is absolutely not to be surpassed. As an engineer myself, I am constantly working in the hope of making this kind of product.

The second joy belongs to the person who sells the product. Our company is a manufacturer. The products made by our company pass into the possession of the various people who have a demand for them, through the cooperation and efforts of all our agents and dealers. In this situation, when the product is of high quality, its performance is superior, and its price is reasonable, then it goes without saying that the people who engage in selling it will experience joy. Good, inexpensive items will always find a welcome. What sells well generates profits, as well as pride and happiness in handling those items. A manufacturer of products that do not bring this joy to people who sell those products is disqualified from being a manufacturer worthy of the name.

The third, the joy of the person who buys the product, is the fairest determiner of the products value. It is neither the manufacturer nor the dealer that best knows the value of the product and passes final judgment on it. Rather, it is none other than the purchaser who uses the product in his daily life. There is happiness in thinking, "Oh, I'm so glad I bought this." This joy is the garland that is placed upon the products value. I am quietly confident that the value of our company's products is well advertised by those products themselves. This is because I believe that they give joy to the people who buy them.

The Three Joys form our company's motto. I am devoting all my strength in order to bring them to reality. It is my hope that all of you, as employees of the company, will exert every effort so that you never betray this motto. I also hope that our agents will understand my desires in this regard so that we may continue to benefit from cooperation.'

The following year was to see the opening of a new chapter in the history of the company.

▲ Close up of Honda's first 4-stroke engine.

Top right: Dream E-Type engine designer Kiyoshi Kawashima ▲ riding the bike again at Suzuka Circuit, Japan 41 years later!

1952

There were three memorable occasions for the Honda Company in 1952. The first was the launch of the Cub F-Type engine kit; the second was the start of exporting complete motorcycles and the third was Soichiro Honda's business visit to the USA to observe American industry. This also led to him purchasing a substantial inventory of production line machinery.

Whilst 4-stroke models were now an important development for Honda, Soichiro accepted that a low cost product was still required for general daily use by Japanese commuters. After much consideration it was agreed to revert to a clip-on engine for the pedal cycle, which by definition required low cost 2-stroke technology. A product that could be easily purchased, easily fitted to a customers pedal cycle and easily affordable.

The all-new product was launched in June as the Cub F-Type, a clip-on 50cc 2-stroke engine with red painted engine covers and a white painted fuel tank. The product was much advanced on Honda's earlier clip-on cycle engines and employed die-cast production processes, something Soichiro was keen to adopt as quickly as possible to achieve high volume production. The product became known as the 'red engine with white fuel tank'. The engine, fuel tank and fitting kit was packaged in a branded cardboard box for easy transport, storage and smart presentation.

It was Fujisawa's dream and his task to mass market this engine kit and use the product to expand the company's dealer network across Japan. At the time around 300 motorcycle dealers existed throughout Japan and Honda only had 20% of these selling their products.

Fujisawa had realised there was a completely undeveloped distribution network that everyone else had ignored. He was thinking of the 50,000 bicycle shops that one could find all over Japan and his plan was to contact all of them by direct-mail. A relatively easy computer task today, but in 1952 this job had to be done by hand!

▲ A Cub F-Type bolt on engine fitted to a Miyata pedal cycle. The photo shows an original 1953 working example on display within The David Silver Honda Collection being demonstrated by three times World Champion for Honda, Freddie Spencer. Further detailed history and the background to the refurbishment of our Cub F-Type is on display within the museum.

Fujisawa wrote a masterpiece of a direct-mail brochure and all available staff wrote out the 50,000 addresses by hand. Over 30,000 cycle dealers responded enthusiastically to sell the 25,000 Yen (£15.00 British pounds at the time!) retail Cub F-Type engine. The initial demand was more than the factory could cope with but they were soon producing 7,000 units per month, taking 70% market share.

Promotion of the new product included obtaining a fleet of light aircraft to shower

needed to start exporting in order to gain global presence and further increase production volume. Starting with the most local countries, export of the Dream E-Type and Cub F-Type commenced to the Philippines and Taiwan.

By Autumn, Honda were the top motorcycle producers of Japan. Sales of both the Dream E-Type and the Cub F-Type were highly successful, but Soichiro remained dissatisfied with the level of precision of his components. Although he wanted to be number one in the world, he realised he would never be able to make a breakthrough using his existing machine tools. So the company decided on a program to invest 450 million yen in the latest machine tools.

The Dream E-Type became the top-selling Japanese motorcycle, notwithstanding the fact that it didn't come up to international standards. Just as Soichiro had said, its performance was 'truly shameful' by comparison with overseas models. Around this time, the import to Japan of European and American motorcycles was resumed and the difference between their bikes and Honda's could be clearly seen. The Japanese bikes always came off second best when tested against the imports. It had seemed such an achievement when the Dream E-Type had made it over the Hakone Pass but that wasn't even a challenge for the imports!

The answer was in engineering and production line techniques with both requiring enhanced machine tools. In November, Soichiro embarked on a special trip to America to see the US industry first hand and to buy machine tools. He also went on site visits to view American automobile mass-production facilities, which he had previously only heard about. At the same time, Kiyoshi Kawashima went to Europe on a similar mission and between them the 450 million yen tooling budget was soon spent!

▲ Mr Honda says goodbye to his family prior to the long flight and trip to study USA manufacturing.

◄ Cub F-Type - popular with the Japanese ladies of the 1950s!

leaflets from the sky all over Japan, promoting both the product and the local dealer. In just one promotion for one new product, Honda had ingeniously created the largest dealer network across Japan, now enabling them to market its forthcoming motorcycle products. This was another example of Soichiro Honda and Takeo Fujisawa's ideal business partnership.

The company had realised that they

1953

Several new models were launched this year including the first of many Benly models. In addition, three new factory plants were established in Saitama and Hamamatsu.

In January, Honda moved its Head Office and its Sales Department to the current site of the Honda Yaesu Building in Tokyo. In addition, the company bought a 100,000 sq.m site in Saitama Prefecture, and then started construction of the Yamato Plant. This is known today as the Wako Plant of the Saitama Factory.

In April, the Shirako Plant was completed and started full-scale production. Sales of the Dream E-Type and Cub F-Type increased dramatically so the company expanded its sales operation with offices in four other major cities: Nagoya, Shikoku, Osaka, and Kyushu.

The slogan '120% Quality' first appeared in an article in the March 1953 issue of 'Honda Monthly'. It was typical of Honda's style. 'When human beings aim for 100% they will always miss by about 1%. If a customer buys one of our products that falls short by that 1%, it will mean that Honda has sold a product that is 100% defective. To eliminate the possibility of missing by 1%, we should aim for 120% quality.' New staff joining the company at that time would get their practical education from Mr. Honda in even more blunt terms!

Soichiro became well known for putting the customer first. He looked at things from the customer's point of view. The slogan 'Aim for 120%' was so effective because it showed how he saw himself as an angry customer who had suffered because of a 1% failure.

Known to his staff, as 'The Old Man', Soichiro Honda would often walk the production lines inspecting his employee's work. He would look angrily at a worker doing something wrong and say, 'you idiot, you fool where do you think your pay is coming from?' Reminding him bluntly that it's from our customers, adding 'do you want to kill them?'. The Old Man was once heard to say, 'A lot of people get angry because they're nice deep down, but I'm not like that. When I see something I don't like I get really nasty, because if there's anything wrong with our products, it could put lives in danger. I just can't tolerate people who don't take their work seriously.' So the person he had got angry with would really feel the pressure, but certainly got the message. The next day, the Old Man would act as if nothing had happened. Afterwards, the workers realised that his anger was his way of educating them.

Also in April, the Cub FII-Type was launched with a 60cc capacity that met the forthcoming Japanese domestic license requirements, further increasing the model range and sales volume.

May saw the delivery of the first consignment of the eagerly awaited new machine tools from the US and Europe. The quality improvement systems, essential to Honda's plans for 120% excellence, were now being put in place.

In July, the Honda Workers Union was formed and by November a new pay structure was introduced with the company implementing its workers system of lifetime employment.

The Dream 6E with a 189cc 3-speed 4-stroke engine.

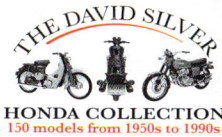

Behind the scenes, a completely new model was under development to be launched in August. It was designed as a practical, low cost, easy to use motorcycle to be known as Benly (from the Japanese word 'Benri' meaning convenient), named by Soichiro Honda himself. The first version was the Benly J-Type and it was the start of many Benly models to follow.

The Benly J-Type featured an 89cc single-cylinder 4-stroke engine producing 3.8bhp. It was based on the German NSU Fox, but the Honda model featured a unique seesaw-type arrangement with the engine attached to the rear swing-arm suspension. This was designed personally by Soichiro.

The idea was that by attaching the engine to the swing-arm and extending its mounting forward, vibration to the rider would be reduced making the ride more comfortable. This was fairly successful, but there were some problems! When the rear wheel went up and down so too did the engine. On a bumpy road the carburettor would shake allowing fuel to leak out affecting the engine's performance. Although the designers aimed at the ideal of 120% quality, it was flawed. However, improvements were made until the Benly earned a reputation as the best practical bike available.

For the Benly, Honda chose a new frame design with a pressed-steel backbone, which was much more suited to mass-production. The pressed frame came in two parts, which fitted together like a sandwich and then welded together. This was how Germany produced bikes at the time, specifically the NSU Fox. Honda engineers considered motorcycles from Germany more advanced than British ones, and many of them had this type of frame.

The Dream model was enhanced to the Dream 3E and Dream 6E models. The 3E was the same as the E-type but upgraded from 2 to 3-speed and from a rigid rear suspension to a plunger design. The 6E had a larger engine with 189cc producing 7bhp with 3-speed transmission.

The company started producing stationary engines used mainly at the time for agricultural purposes. Stationary engines formed a new business division for the company, which would later be separated under the banner of 'Power Equipment'. A division that expanded vastly as various markets opened up for the use of stationary engines.

▲ This wonderfully restored 1954 Benly J-Type is on display in The David Silver Honda Collection with a comprehensive display of historical printed material.

▼ The Saitama factory soon after production started. Today known as the Wako Plant.

1954

A difficult and extremely challenging year, or 'Growing Pains' as Honda would call it!

In January, Honda launched its first scooter, the Juno K-Type. Regarded as the most advanced scooter of the year, making the most of many innovations. At that time Fuji Heavy Industries dominated the domestic scooter market with its Rabbit model followed by Mitsubishi Heavy Industries with its Silver Pigeon model. Scooters were popular and sales were measured separately to motorcycles and auxiliary engines for bicycles. Honda wanted a slice of the scooter sector.

The 200cc Juno K-Type was packed with new features not seen in scooters offered by rival manufacturers. It had the world's first electric-starter and turn-signal lights, fitted to a two-wheeled vehicle. It also had a large all-weather windscreen. It was essentially created as a two-wheel version of the automobile.

Also unique, was the Juno's FRP (Fiber Reinforced Plastic) body panels, a brand-new material for the industry. Soichiro had seen the material used in low volume on the Chevrolet Corvette. Mass-production techniques using this material were however, still under development.

Soichiro had been keen to use plastics for a long time. It would be a first for Japanese manufacturing and the material had to be imported from the US. Honda partly made up the production engineering process themselves, as the procedure was difficult and fraught with many problems. When the material was removed from the mould, it was full of pinholes, pitted and bumpy. When workers polished it, glass fibers would fly off and stick to their skin and when they tried painting it, the paint, as previously used on metal, was not suitable and wouldn't cure. But Soichiro was keen to keep going to find solutions.

Honda enjoyed success in Japan's domestic motorcycle racing

▲ The rare 200cc Juno K-Type scooter as on display in the Honda Collection Hall museum in Japan.

and in January, a party of three Honda staff racers set off from Tokyo Airport for Brazil to take part in the International motorcycle race commemorating São Paulo's fourth centennial. This was the very first time any Japanese riders and Japanese bikes had taken part in motorcycle racing outside of Japan.

The Old Man had told the riders not to expect to win but to finish the race whatever it takes. There was no time to ship the machines by sea and airfreight was far too expensive so the drastic method the three riders used was to take the two motorcycles apart and carry the pieces in their luggage. This was of course in the age of prop planes, and it took them six days to reach São Paulo, but they arrived to a great welcome!

The Honda models were very underpowered compared to the European brands but they finished the 125cc race, as their President had wished, with an average speed of 115kph and placed 13th out of 25 starters, no disgrace!

Pleased with their race result, Soichiro Honda publicly announced the company's declaration of entry in the world famous Isle of Man TT Races. He had been dreaming of an entry in this race for some time and in June he made a special visit to the UK and the famous island to witness for himself, the worlds toughest motorcycle races. It was then, he painfully realised what he had committed his company to!

Soichiro didn't see racing for the fun side, he could clearly see that success in international racing would bring world recognition, of his unknown brand. This would lead to high volume production with export sales. It was a platform opportunity for marketing and Fujisawa could only agree with him.

The ambitious declaration stunned the Honda management and staff. The announcement stated that Honda would enter as early as the following year but Honda was soon to face an unparalleled management crisis. So it was 1959, before the Honda racing team finally made their trip to the Isle of Man.

Fujisawa was a big fan of the Juno and launched a large flamboyant publicity campaign, but it wasn't long before they found that the Juno was a huge failure! The scooter had a series of problems that were not easy to fix. At the same time, sales of the Cub F-Type suddenly came to a halt. The Dream 4E, which had been available as a 220cc from the previous December, started experiencing engine problems. The Benly, after high mileage, was starting to get a poor reputation for noisy gears and tappets.

Emergency measures were taken, production and sales of some models were stopped and focus was placed on increasing production of the most reliable product, the 189cc Dream 6E. Engineers finally found solutions to some of the Juno and the Dream 4E model problems and along with Soichiro, made personal visits to owner's houses to fix them.

However, the crisis was not over! Shipments of the 450 million yen worth of imported machine tools were finally arriving, but the products that would make full use of them had yet to be created. Income was low due to the big drop in sales but payments for the tools could not wait, so Fujisawa had to carefully negotiate with their regular suppliers to delay their payments. If the majority hadn't agreed and the company's bankers, The Mitsubishi Bank, hadn't supported Fujisawa's separate plea, the financial situation at that time would have brought Honda to closure, it was that close!

After its modifications, which required a carburettor fix, the Dream 4E regained its full performance and by the year end reached the largest production figure of the entire Dream E series. The 138cc Benly JA was launched and the Benly J series grew more popular and became one of Honda's mainstay products, with a production run that lasted five years.

Production of the Cub F-Type ceased as too many problems arose from its installation on third-party bicycles that by now varied so widely in both variety and quality. Problems that Honda alone could not resolve.

The first-generation Juno K-Type continued in production for 18 months and some were exported to North America. But total production was fairly low in comparison to other models at just under 6,000 units. However, the plastics technology that Honda laboured so hard over was not in vain as it would reappear five years later in something very special!

1955

A busy year for Honda with four new models launched. In February, the Benly JB was launched. A 125cc model to further expand the Benly J range. A first year original example of this model, finished in black, is on display within our museum. Fully restored to original specification this is a fine example of Benly model development of the time.

By Spring, the company had launched its successor to the Dream E series with the 350cc Dream SB producing 14.5bhp and the 250cc Dream SA with 10.5bhp. These new models were the first to feature Honda's all-new OHC engines and the first to produce in excess of 10bhp. On display within our museum is an original 1956 example of the Dream SA.

At this time, Japan was witnessing a fierce domestic sales war among motorcycle manufacturers that was literally putting their future existence on the line. Domestic race meetings were flourishing, and all the manufacturers competed in them rigorously as the winners could utilise victory to great effect in their advertising.

In July, the recently debuted SA won its first victory in the Fuji Mountain-Climbing Race, which Honda had never managed to win before. Later in the year was the start of what were popularly known as the Asama Races. Honda entered specially tuned machines based on the 250cc Dream SA in three classes, the 250cc, 350cc, and 500cc. These placed second, first, and first, respectively. But in the 125cc class, the Benly JC was defeated by Yamaha's 2-stroke YA-1, which took first through to fourth places.

Soichiro had travelled to Asama, and witnessed with his own eyes the

◀ Opposite Left:
The Honda Race Team with Soichiro Honda (fourth from left) at the Asama highlands venue for the All Japan Endurance Road Race.

◀ A 1956 production example of the 1955 250cc OHC Dream SA. The design and finish is similar to the 350cc Dream SB.

▲ The 1955 125cc OHV Benly JB as on display within the museum.

▲ The Benly JC56.

defeat in the crucial 250cc and 125cc races. He was both angry and embarrassed as the entrant of the 250cc winner was a past colleague and the winner of the 125cc race was Yamaha who had only just started manufacturing motorcycles that year! He was later to coin the phrase… 'Second place is just the first loser'

The approach to racing at the time was to take production motorcycles, that were designed as working machines, and tune them as much as possible and make them look like racing bikes. Successful modifications were transferred to production and failures to the bin. Soichiro clearly saw racing as a laboratory on wheels and retained that belief for many years to follow.

By the autumn, Honda had taken the lead in annual domestic motorcycle production. At this time, in preparation for the 1956 season, the Benly JB was further enhanced with the launch of the Benly JC56 model. A 1956 example is on display within our collection.

In November, the Honda Racing Team participated in the first All Japan Motorcycle Endurance Road Race and won the Manufacturers Championship in two classes.

The Japanese economy started out in a slump this year, but started to improve rapidly from the autumn. Business conditions were the best they had ever been since the founding of the country's new structure.

1956

Two new models and one updated model were launched. Also, the concept for what was to be an outstanding world-beating all-new model was born.

Firstly, the EJ-type, of which little is known, as it seems it was short lived! It featured a 219cc 4-stroke single-cylinder OHV engine, producing 10bhp and functioned as a basic utility motorcycle.

Secondly, the 246cc Dream ME, which was a more powerful version of the Dream SA. It featured larger 18-inch wheels and leading link front suspension, also its lower center of gravity improved its ride quality to make it even more comfortable.

The Benly was also given updates for the JC57 version launched later this year, ready for the next sales season.

Whilst the above and other existing models kept the factories and sales teams busy, Soichiro Honda and Takeo Fujisawa had a vision for an all-new large volume product. They traveled to Europe together to substantiate their idea. The Cub F-type had now gone out of production, as customers were moving on to a pedal style moped known as a 'Mopet' in Japan. The pair wanted something new and different to the competition for both domestic and export markets.

The concept objectives were that the new model had to have the protection of a scooter; the ride of a motorcycle, the simplicity of a mopet and the reliability and performance only a 4-stroke engine could provide. This was the year that the concept for the 1958 C100 Super Cub was born.

1956 was a momentous milestone in the development of this new model concept. In years to come, this new model would establish a world record for the highest volume production automotive model, with over 80 million units produced. Probably the most significant and important model ever in the Honda range!

Soichiro and Takeo were also cementing their goals for worldwide recognition and sales. A new statement was printed in the Honda Company News publication which was under the heading of Corporate Objectives: 'starting at maintaining an international viewpoint, we are dedicated to supplying products

▲ The rare 1956 219cc OHV EJ-type.

▲ A 1957 dated model of the 1956 Benly JC57 as on display in our museum.

of the highest efficiency, yet at reasonable prices, for worldwide customer satisfaction. In making our company grow, we do not expect only to bring well-being to ourselves as employees and to our stockholders. We seek to provide good products to make our customers happy and to help our affiliated companies prosper. Furthermore, we seek to raise the technological level of Japanese industry and to make a greater contribution to society. These are the purpose for our company's existence'.

Soichiro was also often heard making statements such as: 'It's no good just looking at this little Japan; look at the world'.

However, sometimes Soichiro's ambitious ideas would unravel and problems would occur that led to unreliable products. Also new development models that would be scrapped after months of work. But as one senior member of Honda put it; 'if we hadn't experienced those failures, and if we had only been building that day's products, I think Honda would have been a shrimpy little company that probably would have disappeared. If the Old Man's objective had been to be Number One in Japan, and he had taken a more typical Japanese perspective, I have no doubt that today's Honda would not exist'.

In this year, the company established its first official Service Division and introduced a 12-month warranty for all new models.

▲ The 1956 250cc Dream ME. This photo shows a 1957 model as on display in our museum.

1957

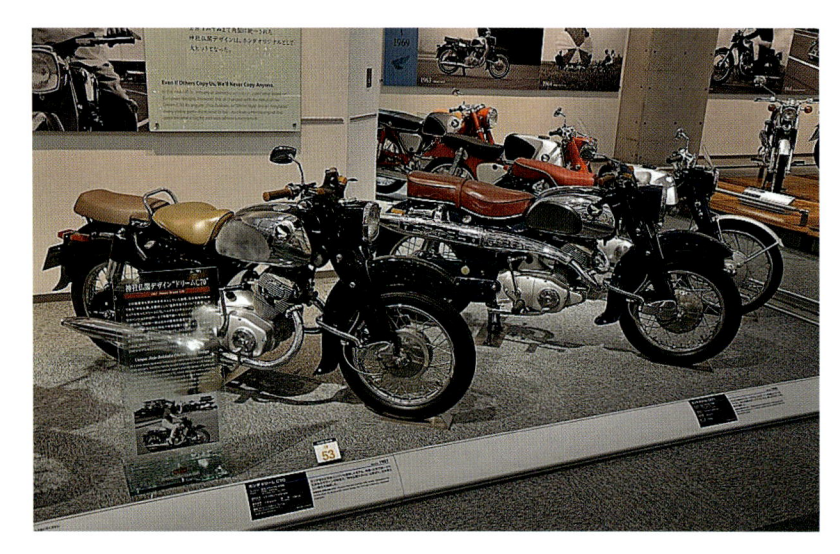

Far left: Soichiro Honda personally testing the new Dream C70.

Left: The all new Dream C70 with its later CS71 sibling, with high level exhausts, as currently on display in the Honda Collection Hall, Japan.

Right: A close up of Honda's first 4-stroke twin cylinder engine as used in the Dream C70.

The Honda Motor Company was by now growing at an amazing pace. Having started in 1948 with the company capitalised at just ¥1 million, in under 10 years it now had a worth of ¥360 million! Later this year the company was listed on the Tokyo Stock Exchange.

They realised from their own past failures, the importance of extensive research and development. A new all-purpose built Honda Engineering Research Centre was established in June as an independent operation. Honda went on to invest more of their profits into R&D than any other automotive company. With the president as the engineering leader it was no real surprise!

Behind the scenes development work was in progress on a revolutionary all-new 4-stroke engine with twin cylinders and in September, Honda's 247cc twin-cylinder Dream C70 was launched. Under Soichiro Honda's personal direction, the new Dream was designed to feature original ideas from the engine through to the cycle parts. The C70 was targeted for export. The shape and style was influenced by the design of Japanese temples to give the bike authentic Japanese originality along with Honda personality. The new model's styling became known as the 'Buddhist temple style'.

A twin-cylinder 250cc 4-stroke was rare at the time as most producers opted for either single cylinder or 2-stroke twins. The Dream C70 already possessed the distinctive Honda character that Soichiro was seeking to express. It was the company's first model to strongly capture that unique Honda personality.

The Dream C70 was also the first product to take full advantage of the powerful new machine tools in which Honda had heavily invested. It was more powerful than Japanese 2-stroke machines in the same class, and it's price was lower. It had a power output of 18bhp at 7,400rpm; this was Honda's first high-rpm, high-power motorcycle.

At the time there was much competition in the industry. As Honda was aiming high with export sales there was criticism from the industry of the new Honda engines ability to last at such high RPM and at such a low model cost! In response Soichiro was scathing…

'We put our emphasis on higher engine power. However, Honda increases power by raising engine revolutions, so some people seem to be criticising our engines for being short-lived. Manufacturers that think this way might not get very long engine life themselves. If the engineering design is poor and precision is low, then the resulting friction will eat up power, so the efficiency is also low. If you try to force higher revolutions on an engine like that, it's bound to break down. But it would be wrong of them to say that, just because our engines would break down, Honda's must break down, too. If Honda engines are so bad, then nobody would be buying them, but the fact is, they are selling very well.

Those manufacturers are using their own level of technology as a yardstick for measuring ours. They should stop that kind of petty behaviour and confess honestly that they aren't able to do what we can do. Only then will they get to the point of realizing that they had better raise their own level of engineering. All they know how to do is suspiciously figure that a 4-stroke twin can't possibly be sold at that low a price, so there must be shoddy work inside it somewhere. People like that aren't experts in the field. When you come down to it, sham expertise is engineering's great enemy.'

Following the new Dream C70, Honda entered its period of originality and creativity. It refused to copy anyone else, even if others copied it. For the following year an additional model, the C71 would feature an electric starter, that the company had gained experience with when making the Juno scooter. A self-starter comes as a matter of course today, but Honda was the first to feature one.

The new tooling machinery that the company had imported was now being used at a ferocious pace and was being operated at speeds that exceeded the manufacturer's recommendations. 'If these were engines, they were being run into the red zone', said one engineer. Naturally, malfunctions and overheating would occur. Honda ignored the rated settings, and so they had to take countermeasures. It was agreed that as Honda had bought the machines outright then they owned them, and should use their intellectual abilities to modify them accordingly.

Using tricks that would amaze the manufacturers, Honda worked the machinery to the limit. For example, they handled overheating problems by attaching a radiator. Also, a single-purpose machine was enhanced into a multi-purpose one. This experience led the company to fabricate its own machine tools. This ultimately resulted in the establishment of the Honda Koki Engineering Factory in 1962.

1958

A significant year in Honda's history. After two years of extensive development, the all-new mass-market model that Soichiro and Takeo had visualised back in 1956, was to be launched.

Never before had so much time, energy and prototypes been dedicated to one new Honda model. The reason being, both Soichiro and Takeo believed they had a new concept motorcycle that could sell worldwide in big volumes. As it turned out, not even they could have foreseen the vast numbers of production units that were to follow!

As previously mentioned, the new model had to have the following features: the protection of a scooter, the ride of a motorcycle and the economy/performance/reliability of a 4-stroke engine.

Every part was new from the engine to the smallest of cycle components. The shape and design was fresh and different to anything else on the market at the time. There were many difficulties in development and production techniques. The engine was the smallest mass-produced 4-stroke engine in existence.

The design team determined that 17inch wheels were the best for ride height and balance but no wheel rim or tyre manufacture produced 17inch products at that time. They were reluctant to tool up for such an unknown quantity. Little did they know it would become the most produced size ever!

The weather protection legshields were required to be made in plastic, to save weight. This enabled Honda to revisit the FRP concepts that they had experimented with back in 1954 with the Juno scooter. However, the material was now switched to polyethylene, which reduced weight even further, and the new model would be the first motorcycle in the world to use a plastic fairing.

The requirement for an automatic clutch was a late request by Soichiro, as he wanted the Tokyo noodle delivery boys to be able to balance their take-away trays on their left hand! The sole initial reason for what became a huge unique selling feature.

During development, Soichiro and Takeo had colossal production quantities in mind. Volumes beyond any quantity of one model ever seen before; to enable this, all factory and sales divisions were gearing up accordingly. An all-new factory at Suzuka was built with the ability to manufacture this model at the rate of up to 50,000 units per month. At this time their top selling model was producing just 3,000 units per month!

◀ A US specification 1961 example of the C100 Super Cub as on display in our museum.

▶ A 1960 example of the 1958 Dream CS71 designed and built for export. A fine example is on display in our Collection.

▲ Honda's first purpose built high-speed test course completed during 1958.

In July, the Honda Super Cub C100 with its 49cc OHV engine was announced. The new model went on sale a month later at a price tag of just ¥55,000 Yen, equivalent to around £55.00 at the time. In fact, when the C100 finally reached the UK shores in 1962, its retail price was just 79 guineas (£79 and 79 shillings (1 shilling= 5p)).

The C100 Super Cub established the motorcycle design term of 'step-thru'. Since its 1958 launch, Honda have now produced over 80 million units, which have been sold in over 150 countries. The world's largest produced automotive product, a record unlikely ever to be beaten!

As you read on, you will see how valuable and significant the C100 Super Cub was to become for Honda. Its originality was a credit to both Soichiro Honda and Takeo Fujisawa, as this one model was to be the source of funding for many leading projects, including racing!

Also new, was the launch of the electric-start Dream C71 and the Dream CS71 with high-level exhaust aimed at the export market. In addition, the Benly C90 was launched featuring a 125cc OHC twin-cylinder engine with 11bhp @ 9,500rpm. The forerunner to the Benly C92 and CB92 models.

In addition to Honda's new R&D facility the company's first purpose built test course was completed near the Arakawa River.

1959

Within just a few months, the Super Cub C100 was in huge demand and production was thriving. Sales were primarily in the domestic market but big plans for export were underway.

While Honda's racing programme on the domestic scene was going well, the company's burgeoning income from their latest products, especially the C100, allowed Soichiro to realise his international racing dreams. So it was finally time to fulfill his earlier commitment and enter the IOM TT Races.

The TT project had been given to a group of young engineers and racers in their twenties. They had actually started the project in late 1958 and had acquired a 1956 125cc single-cylinder Mondial racing machine from Italy. It was their first hands-on experience of an actual European racing bike and from this, the project team constructed Honda's first racing motorcycle, the 125cc twin-cylinder RC141. However, even though they had twin-cylinders, the output was less than the three-year old 16bhp single-cylinder Mondial. However, in time for the IOM they developed the RC142 with over 17bhp and four machines were shipped to the Island.

On the 5th May 1959, the nine members of the Honda racing team arrived on the Isle of Man. That year's TT Races were held on the short Clypse course rather than the long Mountain course. The 125cc class would see ten laps of the 10.92-mile course. The team began learning the course using four of the recently launched road-going 1959 Benly CB92's that the team had also shipped as training bikes.

The first ever team from Japan looked very much out of place on the Island and they had much to learn, not just about the tough Island course but International racing as a whole. However, this young new team took the paddock and the media by surprise finishing 6th, 7th, 8th and 11th positions in the 125cc lightweight class, which won them the prestigious Manufacturers' Team Award. Back in Japan Soichiro Honda was, of course, delighted.

This was the very start of Honda's long and celebrated IOM TT success story.

At this time, Takeo Fujisawa was busy on his international sales strategy. Both he and Soichiro believed that to be number one in the world, you had to first be well established in America, probably the toughest market of all to win over. Despite earlier exports to the US via distributors, discussions with these US importers didn't provide the assurances Takeo was seeking. They laughed at his targets, when asked how many units Honda wanted to sell in the US, Takeo had replied with the figure of 30,000. The American distributors said that would be tough over the year but Takeo said he was targeting that number monthly!

The US motorcycle industry was then selling only around 55,000 units per year. These were large capacity leisure motorcycles as the automobile was the main form of commuter transport. Takeo could see a golden opportunity in an unexploited commuter market, especially for the Super Cub C100 and other forthcoming Honda models.

After much debate, Honda believed they needed to be strong in the US so they should go it alone. By September, American Honda Motor Company was established and trading in Los Angeles, California from a small ex photographic studio building. From here, the company's roots into the US market began and a huge challenge was to follow!

Back in Japan, R&D engineers had not been slow at work as even more new models were on the drawing boards. Launched earlier this year were a range of 125cc OHC twins, based around the previously launched Benly C90. This included the Benly C92, the Benly CB92 Super Sports and the Benly CS92 models.

The new Benly range became significant models for both domestic and export sales. The CB92 was the first model to feature the now famous 'CB' model title and since their recent success in the Isle of Man, they could promote the CB92 as the production version of their TT racer.

Also made available this year was the Dream CE71 Super Sport, primarily for the USA market, but some say that a small quantity went to Europe? It complemented the previously launched Dream C71 and Dream CS71, but the styling was similar to the new CB92. This 250cc twin, that featured a dry-sump engine arrangement, was only produced for a very short period and according to the parts books, only 390 units were made. It is rumoured that American Honda was set to buy back all the CE71's, for which the reason was unknown, but a small number still exist today. A fine example reached $28,000 during bidding at a 2016 Las Vegas auction, but didn't sell!

◀ The 1959 Dream CE71 Super Sport as on display in our collection. Probably the rarest production Honda motorcycle in existence?

▲ The first Honda IOM TT Team in 1959.

◀ American Honda first office established in California.

43

The 1960s - 1960

Honda's rapid growth continued with huge investment in a new Head Office in Tokyo and two new factories at Suzuka and Saitama.

In addition, and important to the company strategy for the future, the Honda Engineering Research Centre had a further expansion and was re-established as Honda R&D Co. Ltd. It had been long debated that R&D should be a separate entity allowing the operation to develop and research its ideas away from the pressures of factory operations. It was Fujisawa's concept and he in particular was extremely keen for this to happen. He even stated later that if his proposal had not been accepted he would have resigned. It is long believed that the 1960 independence of R&D was one of the great success stories of the company's business structure. Honda R&D went on to become admired the world over by its competitors and other industry producers.

During my time with Honda UK in the 1970s and 1980s, I (Roger Etcell, author) was privileged to visit the Honda R&D facilities in Japan on many occasions. It was always a breathtaking experience. I recall clearly seeing so many different designs of engines being put through their paces on test-beds. Many that didn't make production for various reasons and many that did. At the time I didn't believe there was a possible engine configuration that hadn't been produced as a prototype and run by Honda R&D!

New models were in abundance this year, with the launch of the 250cc pairing of the Dream C72 and the Dream CB72 Super Sports. Also there were similar 305cc versions in the Dream C77 and CB77. In addition, the electric-start Super Cub C102; the C110 Sports Cub and a further attempt with the Juno scooter in a 125cc M80 model.

The Dream C72 was launched in February as a further development of the previous Dream C70/71 touring models. Continuing to use the regular pressed-steel frame construction, the C72 now featured an all-alloy engine with a wet-sump arrangement.

▲ The 250cc Honda Dream CB72 Super Sports.

◄ The new independent Honda R&D facility that was admired by many of Honda's competitors.

There was also a Dream C77 (305cc) model available and both became the company's largest capacity models for export to various countries. They featured pressed-steel handlebars and for the following year special versions were produced for the US market known as CA72 and CA77. These featured high-rise tubular handlebars finished in chrome. Later, there were also versions of these models known as the C78/CA78, but visually there was little difference.

Whilst some new Honda models had made their way to England via various means over the past few years, Honda motorcycles were from this year more formally established in the UK and Ireland as they sold their range of models via four appointed distributors. Maico (Great Britain) Ltd based in London for the Benly and Dream models; Scootamatic Ltd based in Nottingham for the Honda 50 models; Artie Bell Motorcycles in Belfast, Northern Ireland for all models and Reg Armstrong Motorcycles Ltd in Dublin, Republic of Ireland for all models. These were the first formal distributors for the UK and Ireland, as appointed by Honda Head Office in Japan.

The first formal model range on sale in the UK during this year consisted of the C100, C102, Benly CB92, the Dream C72 and C77. Some early deliveries also included the C71 and C76 Dream models.

As a point of interest, the Super Cub C100 was more simply known as the Honda 50 (C100) for the UK and many other markets as Triumph Motorcycles held the 'Cub' naming rights at this time.

However, when these new Honda models went on sale in the UK, post-WW2 anti-Japanese sentiment was still widespread and manufacturers such as BSA and Triumph endeavoured to convince dealers not to sell Japanese bikes. Also, the style of the C72/77 was considered to be somewhat unusual and quite old-fashioned in appearance, so early Honda sales in the UK were not as good as anticipated.

That was all to change when R&D reacted to the 'old-fashioned style' comment. The real highlight model for this year was launched in November - the all-new Dream CB72 Super Sports, later regarded as Honda's first true sports bike.

The 250cc Dream CB72 and later 305cc CB77 were the first Honda models to use a race-style tubular steel frame which combined with telescopic front forks and twin-leading front and rear brakes made these models a huge sales success around the world. The styling was modern and aggressive, the bike lightweight and for a 250cc twin, almost as powerful as a British 650cc!

Making good use of the mass-produced OHV

▲ A slightly later version of the 1960 Sports Cub C110 as on display in our museum.

50cc engine from the C100, a sports 50 was launched in the Sports Cub C110 as well as an electric-start version of the C100, the Super Cub C102. Separately, a Juno M80 125cc scooter was also launched but mainly sold in Japan.

The Honda race team continued with their IOM TT programme this year, respectably achieving 4th, 5th and 6th positions in the 250cc Lightweight Race and 6th through to 10th places in the 125cc Ultra-Lightweight Race. This year's competition was even more challenging than the previous year as the full 37¾ mile Mountain course was used for both races.

1961

The company's capital was set at over ¥8 billion, up from only ¥5 million in just 10 years! Honda is now one of the fastest growing companies.

American Honda had now been established for just under two years and it was time for Fujisawa's team to be as focused in Europe. A European Honda office was established in Hamburg, West Germany, to develop sales across the main European markets.

In June, less than three years after its launch, the Super Cub reached 1 million units in production. An incredible achievement from such an all-new design concept. The Super Cub was by now on sale across most of the world.

Honda's motorcycle production was now at over 1.2 million units per year compared to Yamaha at c.150,000, Suzuki at c.160,000 and Kawasaki, who only started the previous year, at around 10,000 units. All British manufacturers between them produced around 250,000 units in this year!

Sales promotion from racing participation was a key strategy for Honda. Both Soichiro and Takeo firmly believed that success in racing would lead to brand awareness and acceptance and this strategy would work the world over. International racing was still very much on Soichiro Honda's agenda and the company's current sales boom provided the funds for Honda to now take on the world in the Grand Prix Championships.

The Honda racing team with Britain's Mike Hailwood

◀ Inside the extensive Honda factories of the early 1960s.

The 55cc Sports Cub C115 ▶ launched in September of this year.

◀ The Honda race team at the 1961 IOM TT races with Mike Hailwood on No.7.

The 125cc Benly CYB92 as ▶ on display within The David Silver Honda Collection.

won the IOM TT races in both the 125cc with the 2RC143 model and 250cc classes with the incredible 4-cylinder DOHC RC162, which produced 45bhp @ 14,000rpm! The team continued their success with overall wins in the same classes of the 1961 World Motorcycle Grand Prix.

From their astonishingly quick international racing successes, the Honda name was on the lips of motorcyclists throughout the world and existing motorcycle dealers queued for the Honda franchise.

Honda's purpose built racing bikes of the early 1960s, were highly sophisticated for the time and they became the testing laboratory for the production road models of the future. The best example came later in the 1960s with the four-cylinder CB750, which started its development in 1966 and launched in 1968.

It's true to say that the success of the C100 during this period provided the funds to go racing which in itself led to the development of models such as the CB750, so without the Super Cub C100 we may never have seen the CB750!

New models launched this year include the 55cc pairing of the Super Cub C105 and the Sports Cub C115 followed by the 250cc Dream CBM72.

The first of the Honda monkey bike models was also produced but not put on sale! The Monkey Z100 featured a strange triangle shaped fuel tank in white and the engine came from the OHV C100. A small quantity were produced but they were destined only for fun-fair rides and not road use. Mainstream production did not commence until 1963.

1961 models on display in our museum include the 125cc twin Benly CYB92, a CB92 fitted with factory racing components under the 'Y' naming, and a Dream CYB77; the 305cc CB77 twin similarly kitted out with original factory race parts.

1962

Following incentives from the Japanese government and Honda motorcycle dealers requesting the company to make cars, Honda made a bold announcement to start production of automobiles. The S360 and S500 two-seater sports cars and the T360 lightweight truck were launched.

Considerable time was spent on the new automobile project, with senior management and Honda R&D temporarily more focused on four wheels than two! Interestingly, Soichiro only wanted to produce his sports cars in two colour options, red or white, the two colours that were at the time banned by the government, as they didn't want automobile colours to be confused with emergency vehicles. Honda protested and their new sports cars were launched in red and white.

Two-wheel development did continue however and a new Honda factory and offices were established in Belgium to assemble and market mopeds for Europe. In many European countries, pedal-cycles were sold in abundance and the opportunity to encourage riders to use motorised mopeds was felt to be relatively easy. Local production with local employment was key to Honda's strategy.

Back in Japan, Honda had learnt much over the past years from making their modifications to tooling purchased from overseas suppliers. They launched Honda Engineering to design and produce manufacturing machinery to its own specification.

Not so many new models were announced this year whilst R&D focused on automobiles, but a 170cc version of the scooter was launched as the Juno M85, followed by a 50cc Port-Cub C240 utility bike.

The Port-Cub C240, also known as the 50 Light, is an extremely rare model built in relatively small numbers. The idea was to offer a cheaper alternative to the regular C100 Super Cub. With just 2-speeds and less equipment than the C100, the C240 didn't capture the results hoped as buyers preferred to pay that little extra for the full C100 experience. The C240 was only sold in Asian markets and was dropped from the range after this year. We have one of just two known C240s in the UK on display in the museum, an example that came to us via an RAF Squadron Leader based at the time in Singapore. The full story is illustrated alongside the display bike.

◀ The 50cc Port-Cub C240.

For the US market, Honda created a new style of model in a 'CL' on/off road category and launched the Dream CL72 Scrambler. Based on the CB72, the 250cc road going scrambler featured twin high-level exhausts, a 19-inch front wheel and high crossbar handlebars. The new model was well received throughout North America and started a range of CL models with different capacity versions to follow. Honda went on to become masters at creating new model ranges from the same engine base.

Meanwhile, the company had for some time been planning its own motor racing venue and circuit. The Suzuka Circuit (later to be known as Hondaland) was completed and in November the All Japan Road Race were held there with Honda winning the 50cc, 125cc, 250cc and 350cc classes. Car racing also became a popular sport at the new venue.

The Honda racing team continued to dominate the world circuits with further wins at the IOM TT races and overall championship wins in the 125cc 250cc and 350cc classes of the 1962 World Motorcycle Grand Prix.

◄ Honda race team line up at the 1962 French Grand Prix.

◄ Car races at the Suzuka Circuit Hondaland in 1962.

1963

The Belgium factory, the company's first outside Japan, had proved challenging to establish, with much to learn about transferring Japanese style business and production practices to a region not accustomed to Japanese companies. In addition, a bad winter had hindered the building project and the planned schedule.

Belgium had been chosen for its fairly central location within Europe and for its good reputation for manufacturing automotive products with a number of small factories established. By May, the production line was up and running, despite certain parts of the building still under construction. The first motorcycle came off the line, a C100 Super Cub. This was a sub-assembly model with some parts imported from the Japanese factories.

Later in the year, Honda designed an all-new moped exclusively for the Belgium factory, to be sold only in Europe. This was aimed at the more relaxed moped laws of the time, which even allowed 14-year olds to ride in some countries. The Honda C310 Moped was the first complete model to be produced outside Japan. It was based on the C100 Super Cub but it was actually very different with moped-style pedals, up-front fuel tank and high handlebars.

Other new models launched this year included the 90cc Benly C200 and a range of 305cc models based on the C72. These were the Dream C77, Dream C78 and the Dream CB77 Super Sports.

Honda moved into the police supply market and produced their first formal police bike based on the CB77, the Dream CP77. This was the start of a range of police models, primarily

◀ By the end of this year the Honda Belgium factory was producing the all-new C310 Moped.

◀ The 90cc Benly C200 as on display.

The Monkey CZ100 Initially sold with a white fuel tank but replaced with a red tank finish from the following year.

The 300cc Dream CP77 Police bike.

used by the Japan Police Force, but also exported to other markets.

Based on the 1961 Z100, which didn't make it to mass production, the Monkey CZ100 was launched for sale this year and went on to become a big hit in many countries including the UK. A unique mini fun-bike adapted from the Super Cub C100 with its tank and seat from the Sports Cub C111. Easily transported in the boot of a large saloon car, the little Monkey bike was bought by many new users including celebrities such as members of The Beatles. It became the start of many Monkey models to be produced by Honda over many years to come. To find an original early production Honda Monkey CZ100 with an original white tank is a rare discovery today!

The final new model this year was the C105H Trail also known as the Hunter Cub, which was an increased in engine size replacement from the previous C100H/C100T Trail 50. This was on/off road trail bike designed primarily for the US market. Based on the Super Cub 55, it featured off-road knobby tyres, a large rear sprocket for hill climbing, a high-level exhaust and a 55cc engine with the same 3-speed semi-automatic transmission.

In September, the 15th anniversary of the Honda Motor Company was celebrated in Kyoto.

In November, the world motorcycle championship races are held at the new Suzuka Circuit for the first time. Honda won the 50cc, 250cc and 350cc classes and went on to take the world titles in the 250cc and 350cc classes having again won the same categories at this years IOM TT races.

1964

Just two years after starting automobile production, Honda announced participation in F1 car racing and became the first Japanese company to do so. The company by now also had a fast growing range of power equipment products in the form of agricultural tillers and general-purpose engines and this year launched their first outboard marine engine.

International growth remained high on the company agenda and Honda offices were established in Paris. Asia Honda Motor Company was also formed in Thailand and new Honda motorcycle production facilities were established in Pakistan and in China through technical collaborations with existing companies in those regions.

New models launched this year included additions to the Benly range with the Benly CS90, the Benly CS65 and the Benly CB125. Also, the Sport 90, Super Cub C65, Super Cub CM90 and Sport CB160 models are announced as well as the CT200 and Dream CA78.

The Benly CB125 featured an OHC twin-cylinder engine and despite its Benly naming (normally allocated to utility style models) it was a sports bike producing 15bhp @ 10,500rpm. It incorporated technology taken from Honda's world GP racers and delivered class-leading performance at the time. The Sport CB160 was very similar but it now produced 16.5bhp.

The Honda Sport 90 (Super 90 or S90) were a 90 cc Honda motorcycle based on the Super Cub and produced from 1964 to 1969. Its single-cylinder OHC air-cooled engine was linked to a four speed manual transmission. The top speed was 64 mph, and the engine was rated at 8bhp. There are a variety of models based around the Honda S90.

Honda 50 (Super Cub) models are by now selling in their thousands in many countries across the globe. As a low purchase price, reliable and economical means of personal transportation, they provided daily transport for people from all walks of life, just as Soichiro and Takeo had visualised back in 1956. The Honda 50 became a workhorse for some; especially Asian markets. It was transport for daily commuters

◀ The Super Cub as often seen put to heavy work across Asia from the 1960s and still today!

◀ The 1964 Sport CB160 twin-cylinder model as on display in our museum.

52

and for others working unsociable hours, such as factory workers, nurses and many others.

American Honda sold 40,000 motorcycles via their 750-dealer network in 1962 and for 1963 the President of American Honda had set an astonishing target of 200,000 units. To achieve this he required something new in advertising. Working with one of the top US advertising agencies, a campaign known as 'You Meet The Nicest People On A Honda' was created and developed through 1963, ready for a huge TV advertising campaign from April of this year. The commercial featured the Honda 50 (C100 Super Cub) and the response was simply overwhelming. People everywhere were clamouring to start their own Honda dealerships and large US corporations wanted to use the Honda 50 as a product in their sales promotion campaigns.

Almost overnight, the Honda 50 became a huge sales hit and established itself as a vehicle for daily activities. It also helped to erase the then deeply rooted motorcycle image of rebellion and discontent. By the end of the 1960s, American Honda annual sales were exceeding 500,000 units.

The 'You Meet The Nicest People On A Honda' ad campaign went on to be used by other Honda subsidiaries around the world, including Honda UK during the 1970s. What was regarded as one of the most successful ad campaigns in any industry is still used today as a marketing case study.

Also in 1964, the Elvis Presley film 'Viva Las Vegas ' in which Elvis and co-star Ann-Margret rode Honda motorcycles (C110 and C100) arrived for screening in Britain (renamed 'Love in Las Vegas') and was promoted throughout ABC Cinemas via the Honda UK dealer network. Elvis was obviously a big Honda fan as he also rode a CB77 Super Hawk in another film of this year, called 'Roustabout'.

The Honda racing team continued to dominate the IOM TT races with wins in the 125cc, 250cc and 350cc classes. They also went on to take the crown in the 125cc and 350cc World Championships.

New for 1964 models on display in our museum include US specification S65 Sport, S90, CT200, Sport CB160 and Dream CA78.

▲ One of the world's most successful advertising campaigns.

1965

Solely produced for the North American market, the CA160 Touring was launched. It followed the now popular twin cylinder SOHC engine configuration as launched on the Benly twins in 1959. The frame and styling reflected the C72/C77 range with a pressed steel chassis, high handlebars and white-wall tyres. With electric start, dual seat and fully enclosed rear drive chain, the CA160 was well equipped for it's time and produced 13.5bhp @ 9,500rpm. The example on display in our museum is a 1966 model.

Up to this year, Honda's largest capacity motorcycle was the 305cc twin, which was based on the 250cc CB72. With the motorcycle manufacturers of the west believing Japan would only produce small capacity motorcycles, Honda dropped a bombshell in April with the launch of the Honda Dream CB450. An all-new design big motorcycle with a new 444cc DOHC twin-cylinder motor providing a top speed of 110mph (180kph).

Known as the 'Black Bomber' in most countries and 'Dragon' or 'Hellcat' in some other markets, Honda's new big-bike was full of new highly innovated engineering for its time. As well as reliable 12-volt electrics and electric-start, the most radical feature was its DOHC cylinder head featuring a unique valve operation. Instead of the conventional coil springs, it used

◀ Honda's first big motorcycle, the Dream CB450 Super Sports (aka The Black Bomber) as on display in our museum.

◀ The Honda E300 Portable Generator.

'torsion bars', rods of steel that twisted to provide the spring effect.

In the UK, the CB450 was first shown and demonstrated at the Brighton Speed Trials in September and then exhibited at the National Motor Cycle Show held for the first time this year in Brighton. During the demonstration run, using the only sample just arrived from Japan and not run-in, the late Allan Robinson MBE recorded a standing-start kilometer time of 30.1 seconds with a terminal speed of 100mph (160kph).

The CB450 went on sale in the UK in February 1966 at a retail price of just £360.00, which was about the same price as a traditional British 650/750 twin! Honda planned a further publicity event by entering Mike Hailwood as one of the riders in the Motor Cycle 500 mile production race at Brands Hatch during July 1966. However, Mike the Bike was only able to complete demonstration laps on the new models as the FIM suddenly deemed the model not classified as a production bike, as it had twin overhead camshafts!

There is a range of CB450 models on display in our museum which includes an original CB450-K0, a 1967 CB450 Police model (CP450), a 1968 CB450-K1 with 5-speed transmission and a 1970s CB450-K6 with front disc brake. We also have a very rare 1967 CB450D Desert Scrambler, which was based on the CB450-K0 but converted by American Honda and their dealers with factory parts to give a scrambler image, but more on that model in the 1967 section.

Also launched this year was the Benly CD90. Mainly for the Asian markets, this model was a lightweight sports motorcycle using the Super Cub engine enlarged to 90cc but with a manual clutch transmission.

After several years of sales in the UK via various distributors, Honda (UK) Limited is formally established as a subsidiary of Honda Motor Company and located at the very fitting address of Power Road, Chiswick, West London. From here, the UK business remained until the year 2000, when the company relocated to Slough in Berkshire.

By the end of this year total annual Honda motorcycle production reached a staggering 1.46 million units (over 5,000 units per day), up from only 43,000 units per year (165 units per day) just ten years earlier in 1955!

In addition to motorcycles, cars, stationary engines and marine engines the company launched its first portable generator, the E300. There is an original sample on display in our museum.

▲ Honda (UK) Ltd Head Office in Power Road, Chiswick, West London from 1965 to 2000.

1966

New models this year included the 50cc Little Honda P25 (known as P50 in some markets); the 90cc Benly CL90, the CL160 Scrambler and the Super Cub C90.

Developed especially for the female rider, the 'Little Honda' was designed to be ridden like a bicycle. It featured easy-to-operate hand brakes and a pedal-start OHC 49cc engine built into the rear wheel housing. When introduced into the UK market the following year the P50 had a price ticket of just £52-4-9d compared to £79-19-0d for a C100 Honda 50 at the time. The 'Little Honda P50' on display in our museum is a 1967 Japan home market example.

North America had a big appetite for street scramblers in the 1960s and the Honda CL range aimed at this market was increasing year by year. The CL90 model was announced this year and its specification included high-level exhaust and braced handlebars. These were standard features for the CL model range, adding a sleek look as well as function. Due to US local taxation laws that favoured 5bhp motorcycles for 15 year olds, the 8bhp CL90 was dropped by 1969 in favour of the CL70, which makes the 90cc model a rare find today. On display in our museum is a 1967 example of the US specification CL90.

The CL160 Scrambler was also launched for the US market. Based on the 1964 CB160, the CL features the same high-level exhaust and braced handlebars as the other CL's. The example on display in our collection is a first year model.

By 1966 the Super Cub models were updated with OHC alloy cylinder heads rather than the original push-rod OHV operation and the C50 and C90 base models were created. The revised model became the most popular version of the long running Super Cub range.

▲ The highly complex engine of the RC166 as on display at the Honda Collection Hall in Japan.

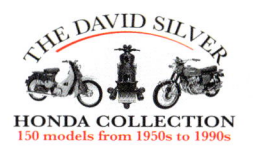

The new CM91 was based on the C90 and fitted with electric start. The new overhead camshaft (OHC) engine produced more power and higher revs.

Thai Honda Manufacturing Company was established in Bangkok to start production of motorcycles for the country.

In April, the company founded Honda Sales Research (HSR) for the pursuit of sales using a more scientific approach. The company would study various issues associated with local sales operations and reflect those findings in the corporation's overall sales strategy. The system brought retail dealers and Honda staff closer together allowing the dealers to feel more involved in future model development.

Having launched the CB450 the previous year, Honda R&D technicians were now looking at the next big model. In particular, American Honda was keen to have something even greater, but what should it be? The next few years were to be spent researching and developing another all-new model that would set the world of motorcycling alight. It would be publically announced in October 1968.

Having won the 250cc and 500cc classes at the IOM TT races, the Honda racing team went on to become the first company ever to win the Manufacturer's Championship in all five classes (50cc, 125cc, 250cc, 350cc and 500cc) of the Road Racing World Championship Grand Prix of 1966.

The 1966 Honda RC166 250cc DOHC 6-cylinder racer was by now one of the most famous works racers ever. First developed in 1964, the evocative sounding 6-cylinder was now producing 60bhp @

▲ The Little Honda P50 as on display in our collection.

▲ The Benly CL90 with high-level exhaust. This 1967 example is on display in our museum.

18,000rpm with a 7-speed transmission and weighing only 112kg. It was an engineering masterpiece and remains a sight and sound even today, some 50 years later! The RC166 won all ten of the 250cc world championship races, with Britain's Mike Hailwood taking the Riders' Championship.

The David Silver Honda Collection does not feature Honda racing machines, as there are just too many types to do their racing category justice! However, it is important to reflect on the engineering achievements of the Honda racing technicians and how much of their engineering success went on to be used in future production machines.

As mentioned previously, Honda used racing as a working laboratory where new technology could be tried before production. Many innovations seen on future road bikes came from their works racers.

1967

Total production of the Super Cub step-thru series of models reached a staggering 5 million units in just 10 years. The Super Cub went on to become the world's largest production motorised vehicle beating the Volkswagen Beetle, Toyota Corolla and Ford Model-T!

New models announced included the Benly SS50, the SS125A, the Benly CL50 and CL125 Scrambler, the Monkey Z50M mini-bike, the PS50K moped, the US only special-edition CB450D Desert Scrambler and finally the UK popular CD175A.

Launched in February, the Benly SS50 was one of the rare 50cc bikes of the time to feature a 5-speed transmission. With 6bhp it boasted the most powerful engine in its class and was designed to appeal to young riders.

The SS125A was a model manufactured for just three years and was based on the Japanese market CS125 along with two other variants, the CD125 and the CL125. All three of these shared many common components. Despite its stylish sporty appearance, the SS125A did lack performance, with only 13bhp compared to the previous 15bhp CB92, in 1971 it was superseded by the CB125. The example on display in our museum is a 1969 US specification model.

Also known as the Benly CL125 in Asian markets, the CL125 was a street-scrambler based on the CB125/CB93 twin. It was the second model in the CL series following the CL72 and came with a slim fuel tank, left-side upswept exhaust and other features. As with many of the CL models the main market was North America where this model sold for two years. It was later replaced by the single-cylinder CL125S in 1973. A 1968 example is on display within our collection.

The Monkey Z50M mini-bike was launched with a similar chassis to the CZ100. The Z50M is probably the most famous of all the Honda Monkey models, it was the first to use the then new OHC 49cc engine. Designed to fit inside a car boot, it featured folding handlebars and a folding seat. The upswept hi-level exhaust and chrome heat shield added to the charming appeal of these very collectable Honda Monkeys.

Featuring a traditional motorcycle style fuel tank and a 3-speed twist-grip gear change, coupled with the reliable overhead valve 4-stroke engine makes the new PS50K a collectable Honda moped. It's rarity means that even rough incomplete examples requiring full restoration attract high values when they occasionally come onto the market, most frequently in France, where the majority of these models were sold! TV presenter James May previously owned the 1969 example on display in our museum.

By mid 1967, stocks of the CB450-K0 were building up in US showrooms. To assist sales of the now fairly dull-looking black only design, Honda produced a special kit of parts for dealers to convert new bikes to a special 'D' models; the 'D' standing for Desert Scrambler. Consisting of tank, seat, mudguards, upswept exhausts and around seventy other parts, to what later became a CL Scrambler style. A 1967 CB450D Desert Scrambler is on display in our museum.

The CD175A was the first of a long line of CD175's that would prove to be a popular low cost commuter motorcycle, especially for the UK. Styling and appearance of this early model was similar to the larger CB450K0 'Black Bomber' but with pressed steel chassis and a twin-cylinder engine based on the CB160, known as the 'sloper' engine. With no electric start, the retail price was kept low and the model often referred to as a 'Benly' has become a collectable model and popular for restoration. In 1967 the UK retail price was just £166-19-0d.

At this time, imported 650cc motorcycles were the largest to be found in Japan, but these bikes accounted for only a small percentage in the overall market. Honda was under pressure to develop a big bike for the US market, bigger than their current CB450. They were not sure how big they should build one, but when they learned that Triumph Motorcycles were developing a high-performance model with a 3-cylinder 750cc engine, the news determined the engine specification. By October, the outline for Honda's new larger model had been defined. The new model was initially targeted at the USA market but had to also be suitable for other markets, including Europe and Japan.

Honda wins continued in the IOM TT races back in June with success in the 250cc, 350cc and 500cc classes. 1967 was an important year in motorcycle Grand Prix history as it was to mark the end of an era, with Honda making the decision to withdraw its racing program from competition. However, Honda went out with a bang, with Mike Hailwood taking the 250cc and 350cc crowns and coming within a whisker of taking the 500cc title.

The PS50K Moped as on display in our collection and as previously owned by TV presenter James May.

The CB450D Desert Scrambler. A rare special as on display in our museum.

The 1968 example of the CL125A Scrambler as featured in our collection.

1968

Having captured five consecutive championship titles in the historic 1966 World Grand Prix Road Racing Series, Honda had made the decision to withdraw from international racing at the beginning of the 1967 season. There were no new championships to win and Honda believed they had learnt a considerable amount of new technology from their racing programme. They now invested that technology in high-performance production machines for customers.

For the past year, Honda R&D, with the aid of their racing teams expertise, had been secretly working on the development of something new for the global big bike market, something that would take the motorcycling world by storm. But now the concept had been decided and in February, a team of twenty members were assembled to design in detail the big new model.

By introducing the new big bike, the company aimed to become the world's top manufacturer in terms of quality as well as volume. The new model's competition would be formidable, since the customer's choice would include comparable models from Triumph, BSA, Norton, BMW, and Harley Davidson. The new Honda therefore would have to offer a superior level of performance and reliability in order to lead the field.

A 4-cylinder, 4-exhaust OHC engine was to be the basis for design, so that motorcyclists in every market could immediately associate the bike with the stunning performance of Honda's Grand Prix racing machines.

The Honda Dream CB750 Four was born. First launched and presented to the public and the world's media at the Tokyo Motor Show in October this year, the CB750 was a revolutionary mass-production 4-cylinder, 4-carburettor, 4-exhaust motorcycle. It also had the world's first disc brake on a large capacity bike.

It was extremely well received by the world's press and quickly became the talking point in motorcycling around the globe. With 12-volt electrics, electric-starter and 5-speed transmission producing a top-speed of 125mph, the new Honda Dream CB750 Four went on to create the new global term of 'Superbike'.

This was the late 1960s, only twenty years after the company had been established and at a time when most motorcycle producers were satisfied with traditional twin-cylinder engine designs. Honda were well ahead of the game and it was their international Grand Prix racing programme that gave them the edge in technology, to not only produce such an astonishing new product, but to produce it in mass volume at an affordable price.

The 3-cylinder Triumph Trident and BSA Rocket-3 with push-rod OHV engine, 4-speed transmission, drum brakes and kick-start only had been launched just a few months earlier. When announced in the UK in 1969, the retail price of the Honda was £695.00, almost the same as the Triumph and BSA triples.

The Dream CB750 Four was the big Honda news of 1968 but there were a number of other new models launched that year.

▲ A 1972 version of the CL350 as on display in our collection.

▲ A 1968 Dream CB250-K0 as on display in our museum. This fine first-year UK example was previously owned by James May and featured on his TV programme, 'Man Lab'.

The Benly CD50 and Benly CD65 models were announced and produced initially for the Japanese market but went on to be sold across Asia. These were sports styled motorcycles using the Super Cub engine with manual transmission.

Also for 1968, a new Dream series of models featuring an all-new twin-cylinder engine and steel-tube frame were launched. Starting with the Dream CB250 and Dream CB350, which became popular mid-range sports bikes in most countries, which were followed by the CL250 and CL350 Street Scrambler versions, mainly for the US markets.

Finally during this year, Honda signed technical collaboration agreements with motorcycle manufacturers in Mexico and Spain.

▲ The Dream CB750 Four. This fine original example on display in our collection is an early 'sandcast' production model with the unique frame/chassis number of 1000750.

1969

The new Dream CB750 Four was presented to 2,000 American dealers in January at a special event in Las Vegas. Honda sales had recently declined in the USA so the launch of the new CB750 was an important presentation to overturn the situation. Soichiro Honda and many Head Office executives attended the function.

At this time, big bikes in North America were selling at over $2,500 so the price announcement of just $1,495 for the Honda CB750 was met with an enthusiastic applause by the attending dealers. It was later greeted with shock-horror by Honda's competitors! Dealers were soon taking customer orders for the new CB750 at a premium!

The same pre-production show models of the CB750 were then shipped to the UK for their unveiling to the British public, dealers and media at the Brighton show in April. Again, the CB750 stole the show and hit the headlines with a positive reception.

From around the world, Honda was soon deluged with orders for the new 750. The earlier production forecast of 1,500 units a year soon became a monthly figure and even that was not enough, so the number jumped to 3,000 units per month!

On the basis of 1,500 units per annum, Honda had planned to manufacture the 4-cylinder crankcases, the most complex of the engine components, using the traditional sandcast moulding procedure, for low volume production. With the urgency to dramatically increase production, die-cast production was called for, so the engineers worked overtime to develop a die-cast mould, not a quick procedure!

First deliveries of the CB750 started in April. For the initial c.7,400 units, the engine cases were sandcast finished but thereafter, from around September, they became die-cast

The mighty 4-cylinder engine of the Dream CB750 Four.

The Little Honda PC50 as on display in our collection.

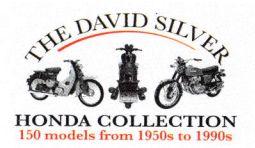

produced. Over 750,000 units of the CB750 SOHC Four in both 'K' and 'F' Super Sport styles were eventually manufactured between 1969 and 1978. While all model types remain collectable, it is the rare sandcast examples that fetch premiums, some selling in excess of £40,000!

In Japan, the launch of the CB750 gave birth to a popular category name known as 'Nanahan', meaning seven and a half!

Within months of the CB750 hitting the showrooms, racing one one of these, was very much on the agenda. The Honda R&D racing team prepared several 750's to compete in the annual Suzuka 10-Hour Endurance Race to be held in August. The CB750 dominated the race with a 1st and 2nd place finish. For America, several CB750's were being prepared for the following year's AMA Daytona 200-Mile Race in March. Veteran rider, Dick Mann streaked to victory, a win that sent customers throughout the USA running to their Honda dealers.

Other new model launches this year included the Dream SL350 twin-cylinder for the USA, the 4-stroke Little Honda PC50 as well as the Dax ST50 and ST70. The Dax models got their name from the resemblance of a Dachshund dog! The handlebars folded down to allow storage in a large car boot.

The Dream SL350 was primarily introduced for the North America market and was the largest displacement model in the SL series. The first K0 model featured a semi-double cradle frame and electric start specification. The example on display in our collection is a 1970/71 K1 model when the frame changed to a full double-cradle design and the electric start was discontinued. The SL range of models went on to be replaced by the new XL range from 1974.

The Little Honda PC50 was based on the Belgian produced C310 model, but this pedal-equipped stepless-transmission model borrowed its simple look from the P25. It was sold in Japan, Europe and the USA and enjoyed wide popularity. With no footrests or kick-start, it is a proper moped with a fully effective cycling capability engaged by the operation of a lever on the right hand crankcase. When the lever is engaged in drive mode, the drive is taken through a single gear and automatic clutch. The 49cc OHV engine moped went on sale in the UK during 1969 at £80.00. The example on display in our collection is a 1969 Japan home-market model.

Launched in late 1969, the Honda ST50 and ST70 mini-bikes were known as the Dax in Japan and Europe. It was fitted with a 4-stroke engine and a 3-speed automatic clutch. Other key features included a pressed-steel 'T-bone' frame that distinguished it from Honda's other mini-bikes, together with fat tyres and folding handlebars. The ST50 went on sale in the UK during this year at a retail price of £135.00 and the ST70 in 1972 at £149.00.

Expanding the export business further, the offices of Honda Australia were established in Melbourne and in Toronto for Honda Canada. Additionally, Honda motorcycle production began in Malaysia through a technical collaboration with a local company.

▲ A later version of the Dax ST70 as on display in our museum. Note the iconic floral seat pattern. On this example the chrome fenders and some other features were added by the customer.

The 1970s - 1970

With the CB750 selling well and creating a whole new image and market position for the brand, Honda launched a huge array of new models for both home and export markets.

Starting with the utility commuter model brand of Benly, the following new models were launched, the majority of which were aimed at the Asian markets but some found their way to Europe and other markets. The Benly CB90, Benly CD70 and then the Benly SL90, Benly SL175 and Benly SL125S plus Benly CB125S, Benly CD125S and Benly CB135 and finally the Benly CL135. In some markets the 'Benly' name was used and in others just the model type such as 'CB125S'.

The new Benly models were a mix of traditional utility design and some sportier in looks to attract a wider customer base. The Benly CB90 incorporated an upright single-cylinder engine, whilst the Benly CD70 featured a pressed steel frame with an engine based on the Super Cub range. The Benly SL90 and SL175 had the on/off road styling for the USA markets and the CB125/135 models were based on a new CB175 design.

In addition to the wide array of Benly models, all-new CB175 and CL175 models were launched with new upright OHC twin-cylinder engines and new tubular steel frames. The on/off road CL175 model with twin high-level exhausts was primarily for the North American markets. The museum features a 1971 version of the CB175, which went on sale in the UK at £269.00

A new CB100 single-cylinder model was launched and exported to most markets including the UK. It featured a simple single-cylinder OHC engine with a 5-speed transmission. This lightweight sporty commuter was popular across most markets including the UK, where it went on sale at just £200.00. Our example of this model in the museum is a 1973 CB100-K2 USA specification type. For the US market, an SL100 on/off road style model was also produced based on the CB100. A 1972 example of the SL100 is on display within our collection.

Finally, for the two-wheel range of new models the QA50 was announced for the North America market only. Featuring a similar OHV 4-stroke engine to the PC50 moped this 'budget' mini Honda was in every respect tiny, yet an adult could easily ride it. This model continued in production until 1975 and is collectable today by Honda mini-bike enthusiasts. The example on display in our museum is a 1974 USA type.

◀ The Honda US90 driven by Sean Connery in the 1971 Bond movie 'Diamonds Are Forever'.

Also launched, was the first of what was to become a large range of Honda ATCs/ATV's (All Terrain Cycles/Vehicles), the Honda US90 (later known as the ATC90) three-wheel balloon tyre recreational vehicle. Initially designed for fun in all-weathers, the Honda ATC's soon found practical use in agriculture and many other fields. This also gave dealers winter products to sell when motorcycle sales traditionally slowed down.

The ATC90 featured a 4-stroke engine from the C90 Cub range. It was first available in North America from 1970, the early K0 model had no lighting but the second K1 model featured lighting and a few other cosmetic changes. The large balloon tyres are no longer made and rare new ones sell today for over $1,000 each!

The US90 (ATC90) was featured in the 1971 Bond movie 'Diamonds Are Forever' with Sean Connery riding the then unusual vehicle in true style! A 1972 example of the ATC90-K1 is on display in our Honda collection.

With production of the 4-cylinder CB750 now in full swing, the Honda engineers at R&D were focusing on developing a surprise smaller capacity version, or two! See the next sections of 1971 and 1972.

Honda launched its annual employees Idea Contest at Suzuka Circuit. Nearly 1,000 ideas were submitted in this first year with the aim of many of the ideas to become reality. The winner of the 'Presidents Award' at this first contest was a compact folding motorcycle, which went on to become the 'NCZ50 Motocompo' model produced from 1981.

Accumulative motorcycle production reached 14 million units this year. In America, the US National Safety Council honored Soichiro Honda for his company's contribution to road safety.

American Honda donated 10,000 mini-bikes to the YMCA.

◀ A 1971 version of the new look CB175 that went on sale in 1970, as on display in our collection.

◀ The styled for USA on/off road SL100 as on display in the museum.

65

1971

Not content with just one 4-cylinder motorcycle Honda engineers had been busy over the past year reducing each component of the CB750 in size to produce a 498cc version and the Honda Dream CB500 Four was launched in April.

In some ways the CB500 Four was regarded by some riders as a better motorcycle than the CB750. The lower, lighter easier to ride 4-cylinder favored many motorcyclists and the performance was still exhilarating with 50bhp. It had a top speed of 115mph and a dry weight 35kg less than the 750, which could be said, made it easier to ride.

Good original examples of the CB500 Four with the genuine Honda four piece exhaust systems in place, are today in high demand. Not quite reaching the prices of the CB750 but as its bigger brother increases in value so does the 500! The CB500 Four went on sale in the UK in 1972 after being launched at the London Racing & Sporting Motorcycle Show in February where it stole the media headlines.

It is not widely known that at the time, whilst the British motorcycle industry was still in decline and now suffering from the launch of the CB750, the then Sales Director of Honda UK deliberately held back the first shipments of the CB500 to arrive in the UK. This was to give the British bike

▲ The Dream CB500 Four as on display in our museum.

▲ The Benly CB50.

▲ A 1972 USA example of the SL125-K1 as launched this year and as on display in our collection.

industry a bit of a fighting chance - a friendly but futile gesture!

By now, as the author of this book, I had owned some smaller Honda motorcycles and larger British bikes, but a 1972 CB500 Four was my first new motorcycle and boy did I enjoy it! It served me well with many trips to the IOM TT, Transatlantic Trophy Races and other regular race meetings across the UK. In 1975 I changed the 500 to a used CB750-K1 and a year later to a new CB750-K6. I was well and truly 4-cylinder hooked!

For this year Honda also launched an updated version of the 750, the CB750-K1 model. With slimmer side panels and re-shaped seat, it was easier to sit astride. The four-carburettors had a new linkage system requiring just a pull and push throttle cable arrangement rather than the previous 1 into 4-cable system. A list of other small modifications further enhanced the CB750 and sales continued to go from strength to strength.

Other new models launched included the Benly CB50, the SL70 and SL125.

The Benly CB50 was first announced for the Japanese market, it was some time before Europe saw these attractive little sports roadsters. Early models had a front drum brake and remained popular throughout Asia for many years. In 1977 a restricted 30mph front-disc brake version of this model known as the CB50J finally made it to European shores. It went on sale in the UK to replace the SS50, as a new sports-moped for sixteeners.

Mainly for the North American markets the SL street/trail range was expanded with the introduction of the SL70 and SL125. Produced for three years, the SL70 featured a 72cc OHC engine and was a favorite with younger riders who used them as a trail bike and as a mini motocrosser. Produced as the K0 model for 1971 and 1972, then the K1 model, with minor cosmetic changes, for 1973. The SL70 was later replaced by the XL70 in 1974. On display in the museum is a first year SL70-K0.

The SL125 was a more serious off-road contender compared to its higher revving little brother, the SL100. A top seller from its launch this year, with its higher torque 122ccc engine and just 91kg overall dry weight. Whilst the SL range of models was created primarily for America, the SL125 was one model in the range that made it to Europe and was sold in the UK until 1974, when it was replaced by the XL125. The SL125 on display in the collection is a 1972 K1 model from the USA.

Total Honda motorcycle production had reached 15 million units. All Honda 4-stroke engines now featured a new valve seat, developed by R&D, to enable the use of un-leaded fuel worldwide.

For the automobile range the company developed and launched a new CVCC engine, the world's first to comply with the forthcoming USA statutory clean air act, something American automobile manufacturers said, was almost impossible!

Honda Motor de Brazil Ltd was established and production of Honda motorcycles began in Mexico.

1972

Testing R&D's engineering boundaries further an even smaller version of the CB750 and CB500 was launched this year - the Dream CB350F Four.

A masterpiece in precision motorcycle production engineering, the CB350F was not known for its top speed but more for its overall lightweight agility. This model was said to be a personal favorite of Soichiro Honda for its smooth and well balanced performance.

Whilst successful in Japan, USA and other markets, the CB350F was not formally imported to the UK. The CB750 and CB500 fours were both selling well in the UK and it later became known that a 400cc version based on the CB350F was also under development with styling more suited to European tastes. So Honda UK held off on the CB350F and waited for the 400 version, which didn't actually arrive until after its launch in late 1974.

A small number of CB350F Fours have however been privately imported to the UK over time and are today, a collectable model for enthusiasts.

The Dream CB350F Four on display in our collection is a 1972 early production example from the USA.

◀ A masterpiece in precision production engineering, the 1972 Dream CB350 Four from our museum display.

◀ The world's first mass-produced 4-valve single cylinder engine as featured in our 1972 XL250 Motosport.

Honda R&D had also been busy with yet another array of new models to suit various markets and customer requirements. Launched this year, was the Benly CB90JX with a front disc brake for the Asian markets. Also, a Chaly CF50 and CF70 semi-scooter based model for the commuter markets and the new ST90 Trailsport for North America. Unlike the ST50 and ST70, the 90 was an on/off road styled fun bike.

Honda also moved further into the on/off road and competition markets. They announced the 4-stroke XL250 Motosport to start replacing the ageing SL series. Also, they added their first production motocross bike, the 2-stroke Elsinore CR250M. These were the start of many XL and CR models to follow.

The 1960s Honda SL range of models had sold well across North America, but there was now a hunger for dual on/off road bikes from other markets especially Europe. So R&D went back to the drawing board to design an all-new 250cc model, the XL250 Motosport. Initially however, it was known as the SL250S in some markets. This was a fully-fledged dual-purpose model with a lightweight frame and an all-new single-cylinder 4-stroke engine featuring a world first 4-valve mass-production cylinder head. The new model sparked a widespread interest in trail-bike riding and led to a range of new XL models in an assortment of engine sizes to follow.

Today, an original XL250-K0 is a desirable collectors bike, the one on display in our museum is a 1973 USA specification example.

Soichiro Honda had previously gone on record saying that his company would never build 2-stroke powered motorcycles again, after their earlier 1940s/50s models. He believed that 4-stroke was the more correct environmentally friendly product. However, competitive motocross racing had moved from 4-stroke to 2-stroke power and Honda was forced to develop a 2-stroke competition bike for the fast-growing sport of motocross. Soichiro finally, but reluctantly, gave his approval for both a 2-stroke works bike and a mass production bike for motocross. The Elsinore CR250M was announced as the production version.

Named the CR250M globally, it was subtitled the Elsinore in North America, after the annual off-road race at Lake Elsinore, California. Popularity of the CR250M rose after a stock machine won the 1973 AMA 250 national motocross series and Steve McQueen would often be seen riding one during the 1973 season.

▲ Honda's first production motocross bike, the 2-stroke Elsinore CR250M as seen in our collection.

1973

Now in their late 60's, President Soichiro Honda and Executive Vice President Takeo Fujisawa stepped down from their roles together to formally retire, both becoming Supreme Advisors. From here on, Soichiro also became an Ambassador for the company, continuing to travel the world to visit the many Honda subsidiaries, factories and related events such as racing. Kiyoshi Kawashima was appointed the company's new president.

This year was Honda's 25th Anniversary that was celebrated with the achievement of a cumulative total of 20 million motorcycles produced. Joining in on the celebrations were top dealers from 37 countries who were selected for a special all-expenses paid 'Silver Jubilee Holiday in Japan'. It was the first time for the lucky dealers to visit the Honda factories and experience the many wonders of a country very rarely visited at the time by westerners.

Honda was by now returning to racing and established the Honda RSC (Racing Service Center Corporation), to allow further development of worldwide racing activity in all categories. Later, the company became HRC (Honda Racing Corporation).

Honda R&D was also very busy continuing to launch new models with the following announced for this season; the Novio PM50, a 2-stroke model for the Asian markets, the

The popular TL125 on/off road trials-bike as on display in our collection.

The rare MT250 road legal motocross specification model, as on display our museum.

TL125 trial-bike, the Elsinore MT125 and Elsinore MT250 plus the CY50, the ATC70 and the CB250T and CB360T twin-cylinder models, known as G5 in Europe.

The new 4-stroke TL125 was launched to take on the 2-stroke trial bikes of Europe. It was Japan's first ever trials bike and was known as the 'Bials' to promote the relatively unknown field of motorcycle trials throughout the country. The engine was based on the SL125 with a 5-speed transmission and re-tuned to just 8bhp @ 8,000rpm. The lightweight low-price competition bike became a favorite for several years and went on sale in the UK in 1975 at a retail price of £329.00.

Following the success of the CR250M launched last year, there was demand in North America for a similar bike for road use and the Elsinore MT250 followed by the Elsinore MT125 were both announced. These models were similar to the CR250M but equipped with lights and a silencer for street use. These new MT's represented Honda's first road-going 2-stroke models since the debut of the Dream D-type, 24 years earlier!

The new 3-wheel product of ATC90 (US90) was a great success, so Honda introduced a 72cc version, a size suitable for junior riders. Debuting this year, the ATC70 featured an OHC engine based on the Super Cub motor with a 3-speed semi-automatic clutch transmission. Whilst mainly produced for the USA markets, the ATCs did arrive later in the UK starting with the ATC110 in 1980.

Originally launched in 1967, the CL125 was an on/off road scrambler design that originally featured a sloper design 124cc OHC twin-cylinder engine. From this year however, the CL125 was re-launched with an all-new design featuring 122cc single-cylinder OHC engine. Almost identical to the CL100, this new single-cylinder model produced more low-end torque than its earlier sibling and had a lower overall weight, which made off-road riding that much easier. The example on display in our collection is a 1974 USA specification CL125-S1.

Adding to the new XL range, which at this time was gradually replacing the CL/SL series, the XL175 first entered the market. It featured a new 173cc single-cylinder OHC engine constructed of magnesium-aluminium for lightweight dual-purpose motorcycle riding, the model continued until 1978. On display in the museum is a 1975 USA specification example.

◄ The new CB250T & CB360T (known as the G5 model in Europe).

The twin-cylinder sports models of 250cc and 350cc took on a completely new look with the launch of the CB250T and CB360T, known as the 'G5' models in some European countries. The two models went on sale in the UK at £539.00 and £599.00 respectively. It was felt these two new models were rushed through R&D as they suffered a few technical problems during their warranty period. This included a recall to have their handlebars changed due to the potential of them cracking under general use! At this time, I was working for a Honda dealer in East London and clearly remember having to change a number of handlebars on some stock and customer bikes!

1974

The year that the UK's most admired Honda model was launched! A popular model with both male and female riders of the time, and today with restorers. The motorcycle that surely has the world's most stylish exhaust system of the 1970s – the CB400F Super Sport.

Based on the previously launched 4-cylinder 347cc 5-speed CB350F, that wasn't formally sold in the UK, the 408cc 6-speed CB400F became available in the UK from early 1975 at £669.00, two thirds of the price of the CB750 at the time.

The lightweight, low seat height compact super-sport model was admired by all ages and soon became a big hit, not only in the UK but also across many other markets including the USA and Japan. It was also a successful production racer and today the CB400F is the most popular model for restoration with our parts department. On display in our museum are examples of both the CB400-F1 and the later F2 version as available from 1977.

Innovation after innovation, Honda was flying high and they again stunned the world of motorcycling, with the launch at the Cologne Motorcycle Show of the amazing GL1000 Gold Wing.

The all-new 999cc Gold Wing set new touring standards with groundbreaking engineering featuring a flat four-cylinder liquid-cooled engine with shaft-drive and a fuel tank under the dual seat. Over 40 years and many versions later, the Gold Wing continues in the range. Gold Wing models were initially produced in Japan from 1975, then due to it's US popularity, in Ohio USA from 1980 to 2010 but today back in Japan. The GL1000-K0 went on sale in the UK during 1975 at a price of £1,600.00. The USA specification model on display in our museum is a rare first year example.

Other new models launched included the CB500T; the CB550 Four an increased CB500-K with a few cosmetic changes; the CL360 dual-purpose bike mainly for the USA; the single-cylinder XL100 and XL350 models to add to the fast growing XL range; a junior MR50 primarily for North America and the twin-cylinder CB200 replacing the previous CB175.

The CB500T was intoduced but ran for just two years. This replaced the long running CB450 twin (later versions not sold in the UK) using the same DOHC

▲ A UK favorite, the CB400F as on display in the museum.

◀ The last of the DOHC twin-cylinder engine models the Honda CB500T as on display in our museum.

parallel twin engine with a larger capacity. It was sold in many markets including the USA and the UK. This model offered a lightweight and low price alternative to the new CB500 four-cylinder model. The CB500T went on sale in the UK during early 1975 at £699.00 compared to £859.00 for the CB500 Four. However, it wasn't a great success.

The CB200 became a popular model for daily commuters in the UK. Having replaced the CB175, the new 198cc twin-cylinder featured a cable-operated front disc brake. Dual carburettors, 5-speed gearbox and electric start remained from the CB175. A distinguishing feature of the CB200 is the rubber trim down the centre of the fuel tank. In some markets it was known as CB200A/B, CB200K or CB200T. The early USA example on display in our collection has a very early VIN number of just 1000513.

This was the time of the sports-moped sales war in the UK. All mopeds still required pedals and Honda's sports-moped offering was the 4-stroke SS50 launched at £210.00. However, it was soon to be up against the 2-stroke models from Yamaha, Suzuki, Puch, Fantic, Gilera and others. Whilst the Honda was the more sensible option, with its quieter and cleaner 4-stroke engine and good looks, 16 year olds wanted the fastest and of course the Honda was the slowest! Things improved a little for Honda in 1976 with the launch of the quicker SS50-ZB2 and today both SS50 models have gained a strong following by collectors and restorers.

Now in his retirement, but still active travelling the world as the company ambassador, Soichiro Honda received an honorary Doctor of Engineering degree this year from the Michigan Technological University.

Honda Suisse was established in Geneva and construction of the Kumamoto factory began in Japan. In addition, the company's Head Office moved to larger premises within Tokyo.

▲ The first of many Gold Wing models, the GL1000 as seen in our collection.

A fine example of the 1974 SS50Z Sports Moped as on display in the museum and as once owned by James May. ▶

73

1975

The 4-cylinder CB750 was continuing to sell well across all markets. By now, North America had the K5 version, small enhancements were made to each new years model in the USA, but Europe continued with the K2 type preparing for a new K6 version due in the following year.

The CB750-F1 Super Sport was introduced as an alternative styling to the 'K' series. Whilst the engine remained similar, the cycle parts were heavily revised to offer a sleek sporty styling. Notable features included a 4 into 1 exhaust system and a rear disc brake. On display in our museum is a 1976 UK registered example as used by its previous owner in the Isle of Man.

To further develop their off-road competition models, Honda R&D had recruited Britain's Sammy Miller in 1974 to help produce a new range of off-road competition bikes. The first result was the successful 1975 TL250 competition trial-bike. Sammy was not only a multi British and European Trials Champion, he was also an accomplished road racer and IOM TT competitor.

The new off-road only production TL250 was launched for worldwide sales. The 4-stroke was a direct development from the factory competition bike, to help reduce weight, magnesium-alloy was used in the engine and to help increase power a 4-valve cylinder head featured in the single-cylinder engine. This model was a serious over the counter competition machine and is sort after today for classic trials competitions. An optional lighting kit was also made available. The example on display in our collection is a 1976 USA type.

For the past two years, a dedicated team of Honda R&D

◀ A 1976 UK registered CB750-F1 as on display in our museum.

◀ The TL250 trial bike that Sammy Miller helped to develop and as on show in our collection.

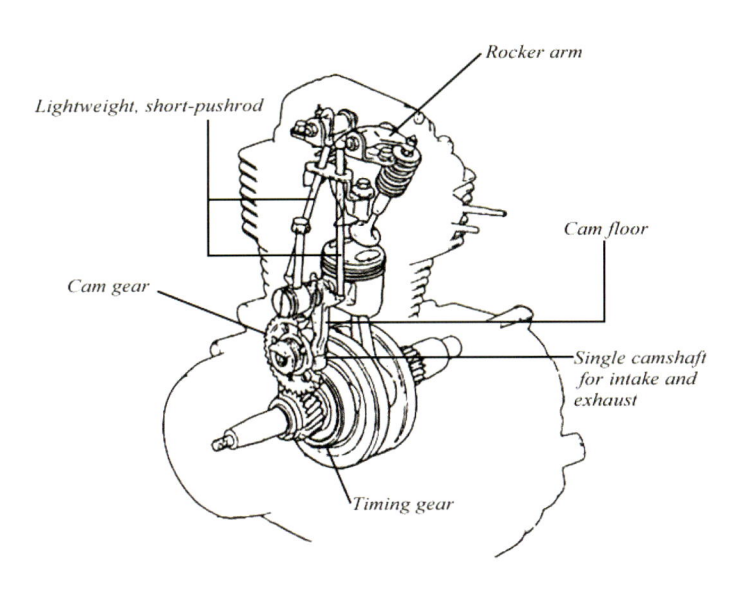

The unique, short-pushrod OHV engine of the CG125 featured a gear-driven, single camshaft structure for both intake and exhaust. The shaft is located where the cam-chain housing is found in a more conventional OHC engine.

engineers had been studying consumer use of Honda and other makes of motorcycles in major cities of developing countries such as Thailand, Indonesia, Iran and Pakistan. Their findings shocked and concerned the team as they found that users approach to their daily transport was very different to that of Japan. As it is today, it was common to see small motorcycles such as the Honda 50 carrying all family members including father, mother, one or two children and the family dog! In addition, many small bikes were seen carrying exceptional loads, including livestock, and even some towing trailers full of even more strange content.

It was also evident that owners in these countries didn't understand or care about routine maintenance and only visited dealers and workshops when their motorcycle stopped working! Many bikes were struggling to run on oil that had turned to goo and paper air filter elements that were solid with dirt! Drive chains were so stretched that they were hitting the chain cases causing damage.

It was concluded that due to their complex structure, the current 4-stroke OHC motorcycle could not perform to it's true potential in developing countries, so Honda therefore decided that they should develop a new lightweight motorcycle especially for these markets. It was to be above all practical, durable and as maintenance-free as possible.

The basic concept of the all-new motorcycle was established. It was to employ a unique 4-stroke OHV lightweight short pushrod engine for higher performance and easier maintenance. The frame design was to feature more than usual strengthening in order to cope with the unusual loads and the air-cleaner element was changed to a new washable material, to withstand repeated cleanings.

The all-new CG125 was announced and ready for production. It was exported to developing countries, where it was extremely well received with exceptional annual growth in sales. The same CG125 also went on sale in the UK from next year at just £299.00.

At this time, R&D was also researching a replacement for the contact breaker points based ignition system as featured at the heart of all their current models. Although not new in the automotive world, Honda was now developing their own CDI (Capacitor Discharge Ignition) system which would feature on many new models from 1977 onwards. This was to provide benefits such as low maintenance and more accurate and reliable ignition timing.

Also announced was the MR175 Elsinore, a 2-stroke Enduro model with road legal lights primarily for the North American markets and all-new versions of the XL125 and XL250.

Production of Honda motorcycles started in Brazil with the launch of Moto Honda da Amazonia Ltd. and their first product was the above new CG125!

1976

To further enhance the CB750 range of models and to assess the interest of motorcyclists for automatic transmission, Honda R&D developed a 2-speed semi-automatic version of their car based Hondamatic transmission for motorcycle use. The CB750A Hondamatic was only produced for the North American market, it was based on the regular CB750 with reduced power and more weight, as the Hondamatic version was designed intentionally to be much slower than its sporty brother.

The 2-speed transmission included a torque converter typical of a regular car automatic transmission but it did not automatically change gear. Each gear is selected by a foot-controlled hydraulic valve/selector, similar in operation to a manual transmission motorcycle. The foot selector controlling the application of high pressure oil to a single clutch pack (one clutch for each gear), causing the selected clutch (and gear) to engage. The selected gear remains selected until changed by the rider, or the kickstand is lowered to shift the transmission into neutral.

The CB750A Hondamatic remained in the range for 3-years, the model on display in our collection is a first year example.

This was the year I joined Honda UK Head Office in Power Road, Chiswick, West London. I started in the Service Department undertaking various tasks and I was then the proud owner of this year's revised CB750, the K6 model. I rode it daily from Romford in Essex to Chiswick, almost the entire North Circular, or often for a change, a ride through central London! As I was regularly doing high mileage I would often take a different new model home from the press fleet to run-in, as was essential in those days! Lucky me! We had received one sample CB750A Hondamatic from Japan to evaluate and I spent a number of days and considerable miles riding what was at the time a strange and intriguing but not necessarily pleasant experience! For the USA with mainly straight roads you could possibly see some benefits but the torque reaction and gearing ratios always seemed to be in the wrong place at the wrong time, often whilst cornering providing a strange and unwelcome sensation. Not suited to European roads, the model remained exclusive to North America! However, the Hondamatic system did reappear in a model destined for the UK in 1978!

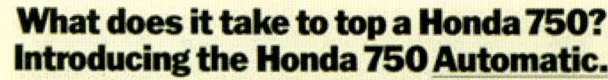

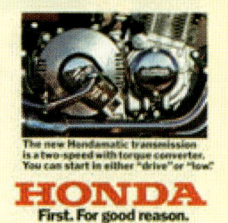

▲ The CB750A Hondamatic as sold in North America for 3-years and as on display in our collection.

▲ The new pairing CJ250T and CJ360T twin-cylinder models.

▲ The Bol d'Or winning Honda RCB1000, based on the CB750 Four.

The ongoing CB250 and CB360 twins were treated to a comprehensive makeover and re-launched this year as the CJ250T and CJ360T, replacing the previous CB250/360 G5 models. What was regarded as the last of the 1960s family of mid-range twins, the new models were an attempt to inject life back into performance. Reverting to a 5-speed transmission and deleting the electric start and even the centre stand, which did cause customer annoyance, reduced the overall weight to aid performance. With styling more appealing to European tastes, the CJ250T and CJ360T went on sale in the UK at £549.00 and £625.00. In North America, where most of the larger version was sold, the 90mph CJ360T became popular for conversion to café racers, and to find an original model today is a rare sight!

What was to become a popular daily commuter model in the UK by the end of this decade, the NC50 Roadpal moped (later to be known as the Express in other markets including the UK) was launched initially for the Japanese market. A new design targeted at women, this model featured 14-inch wheels, low seat position, and a new easy to operate wind-up starting system. It offered all the convenience of a motorbike, coupled with the simplicity of a bicycle. A 1978 example is on display in our collection.

Also launched this year was the CB50J sports moped, the TL50, MR250 Elsinore, the CB125T and a GL1000 Limited Edition Gold Wing.

The CB50J, also known as the Benly CB50JX in some Asian markets, arrived for sale in the UK the following year at just £319.00. It replaced the by now dated SS50 and featured sporty styling without the pedals, now no longer required as part of a forthcoming change in legislation requiring all mopeds to be restricted to a top speed of 30mph.

The TL50 went on sale in limited markets with its stylish lines coming from the larger TL125 and TL250. It was equipped with a 5-speed transmission.

The MR250 Elsinore continued the on-road motocross theme for the USA markets but was only produced for this year. The example on display in our museum is a rare find with a frame number showing it was just number 266 off the production line!

With the 125cc market becoming more important, Honda developed and announced a new CB125T twin-cylinder sports model. This arrived in the UK the following year with a retail price of £489.00.

Finally for this year, a Limited Edition model of the GL1000 Gold Wing was announced. It featured gold-anodized wheel rims with gold colour plated spokes and a small number of other enhancements. The GL1000LTD Edition was just for the USA market and only produced for this year. An example of this model is on display with our other Gold Wings within the museum collection.

Honda racing participated in the European Motorcycle Endurance Race Series. They won the famous Bol d'Or 24-Hour race with British rider Alex George and French rider Claude Chemarin riding the RCB1000, a DOHC 4-cylinder racing motorcycle based on the CB750 Four.

1977

Soichiro Honda and company directors had for some time been concerned about the environmental impact on life surrounding their extensive factory sites. At the time, the Kumamoto Factory alone stood on a site of some 500 acres. Many more acres had been cleared and occupied during the company's wide-ranging expansion of factories during the 1960s and 1970s.

A programme called 'Hometown Forest' had been established, and through this Honda committed to support the planting of trees at all of their factory sites around Japan. In this year alone, over 250,000 small seedlings were planted by factory employees, who went on to care for the protection of the trees, throughout Honda's Kumamoto, Suzuka, Hamamatsu and Saitama factory sites. By 1986, the Kumamoto Factory was infused with greenery covering more than half of it's grounds. The factory was given the Ministry for International Trade and Industry Award for the Promotion of Green Factories, in recognition of its effort to promote planting.

For some time, Honda R&D engineers had been researching motorcycle wheel design in the search for something new to replace the dated conventional spoke wheel. After many experiments, the unique to Honda ComStar wheel was launched and in this first year it featured on several new models.

The move away from conventional spoke wheels was a huge engineering achievement at the time. The new ComStar wheel design provided many benefits: it was lighter than an alternative solid cast wheel, it retained strength and unsprung weight for ride characteristics, it allowed the use of tubeless tyres for enhanced tyre design and it was quicker to produce in the factory compared to lacing a spoke wheel. The design, which continued to evolve over time, featured five pairs of plate spokes bolted to the hub and riveted to an aluminium outer rim.

Today, for restorers the 1970s/80s ComStar wheel remains the subject of many debates about cleaning and dismantling of the assembly. Pressure washing is by far the best method to clean the inside and dismantling is certainly not advocated, as stated on each ComStar wheel!

New models announced this year included Honda's first V-twin, the CX500; new CB250T and CB400T Dream twins; the CR250R Elsinore for North America; the CB750-F2; the CT125 Trail bike and the GL1000 Gold Wing Executive for the UK market only.

The CJ250T and CJ360T launched only in the previous year were replaced with the all-new CB250T and CB400T Dream models, also known as the Hawk in North America. Featuring a new 3-valve per cylinder and counter-balancer (to reduce any vibration) engine plus the new ComStar wheel with a fresh design to tank, panels and seat, the new Dream models were well received. They went on sale in the UK later this year at £729.00 and £829.00 respectively.

The CB750F Super Sport featured an update this year with many new features including the ComStar wheel, a black finish on the engine, a new silencer

▲ Honda's first V-twin, the CX500 as seen in our museum.

and dual front disc brakes. The example on display in our collection is a 1977 USA specification model.

Honda launched the CT125, an on/off road workhorse model, a 124cc 4-stroke engined model primarily designed for agricultural use. Based on the XL125 with lower gearing, solo seat, rear carrier, engine guard and enclosed rear chain case. It featured a 19-inch front wheel compared to 21-inch on the XL. Popular in Australia and other agricultural markets, the CT125 briefly featured in the UK range. We feature a 1977 USA example in our museum display.

As per American Honda, Honda UK also established the idea of a Limited Edition Gold Wing and launched, exclusively for the UK market, the GL1000 Gold Wing Executive. Featuring a

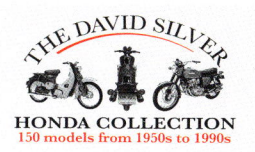

▲ The CB250T Dream and its unique ComStar wheel as seen on display in our collection. ▶

full touring fairing, cast aluminium wheels and a special paint finish the Executive was based on the 1977 K1 model and just 52 units were commissioned by HUK to Rickman Bros. Motorcycles. We have one on display along with it's authenticity certificate, No1 out of 52.

Probably the biggest news of a new Honda model this year was the late 1977 announcement of the CX500 V-twin. Initially named the GL500, this was Honda's first V-twin engine motorcycle and was one of the most advanced engineering projects of the decade.

The all-new CX500 included innovative features and technologies that were uncommon at the time such as liquid-cooling, shaft drive, modular wheels and dual CV-type carburettors for reduced emissions. The complex OHV engine features a crankshaft aligned longitudinally with the axis of the bike and an included cylinder angle of 80° with the cylinder heads twisted at 22°.

The CX500 went on sale in the UK the following year, at a retail price of £1,249.00. It had a fairly lukewarm reception but then went on to become a popular model in the UK especially amongst courier riders who enjoyed it's robust and low-maintenance design. Our display model is a 1978 USA type.

By this year Honda UK had created its own independent racing team, under the leadership of Gerald Davison as 'Honda Britain Racing'. The new team immediately took Honda back to the IOM TT where they won the Formula One class with Phil Read aboard a special 820cc version of the CB750.

The above success led to the launch in 1978 of the Phil Read Replica production model, based on this year's CB750F2, one of which is on display in our collection.

1978

Production of Honda motorcycles reached a landmark with a grand total of 30 million units produced. But we can best start this year with Honda's breathtaking statement, the world's fastest production motorcycle - the CBX!

Seen in prototype form at a Honda press launch in Japan the previous November, the exquisite DOHC 24-valve 6-cylinder superbike was by now the big talking point of the motorcycle media and motorcyclists around the world.

It was designed and developed in just 18 months by Honda R&D's famed engineer, Shoichiro Irimajiri, who had been responsible for the famous 1960s six-cylinder GP racers as raced by Mike Hailwood and Jim Redman. The CBX was not only the flagship model in the Honda range but a testimony of Honda's engineering ability. The CBX is a direct descendant of the 6-cylinder race engines, which is why it took only a year and a half to develop.

The Honda CBX wasn't of course the first 6-cylinder superbike, as Benelli had previously launched their SOHC Sei model which actually was an engine design format copied from Honda's CB500 Four!

Although bulky looking, the CBX engine was only wide at the top. The width across the crankshaft was relatively low as the CBX had its alternator and ignition items stacked behind the cylinder block. This arrangement produced acceptable engine width low down and moved critical items out of harm's way in the event of grounding.

Respected British journalist L. J. K. Setright once wrote of the CBX's width: 'Don't tell me that its engine is too wide: It is no wider than the legs of a rider, so it adds nothing to the frontal area, and personally I would rather have my legs shielded by a cylinder apiece than exposed to every blow'.

Back to my role at Honda UK, by now I was in the Sales & Marketing Department with responsibilities for launching new models, producing sales materials and for looking after the bike press fleet. Each June we supplied bikes to the IOM TT Marshals and for this year we just had to supply them the new CBX as a promotional exercise. The problem was that production was delayed and we would not have stock of the new model arriving in the UK until later this year. Shipments from Japan took at least three months at this time. With the IOM TT as its original proving ground, it didn't take too much persuasion to get Honda Head Office in Tokyo to agree to airfreight four CBX's in time for the TT races. The bikes needed to be in the Island at the latest by Friday prior to practice week. After a number of delays the four crates arrived at Heathrow Thursday and it was that evening before we got them cleared through customs and across to our Chiswick workshops. They needed building from the crates and running-in! Four of us assembled and PDI'd them overnight and in the morning rode them to Liverpool docks, just making the midday ferry for the 4-hour crossing to the Island. Tokyo to the Island with 230 miles on the clock in less than 48hrs! Tight timing but what a ride!

The CBX became available in the UK later this year at a retail price of £2,560.00. We have on display in our collection each of the three versions: the CBX-Z, CBX-A and CBX-B.

Even today, to ride an original CBX is something very special!

It was a big year for new Honda models, not only the CBX but also a pair of new DOHC 4-cylinder bikes in the CB900F sports and CB750K touring. Plus the CB250N and CB400N Super Dreams, the CB400AT Hondamatic and the Benly CM125T.

After ten years of production of the SOHC 4-cylinder CB750, it was time for a fresh model. With technology from the RCB1000 endurance racing bikes and all-new European styling, the DOHC 16-valve 4-cylinder CB900F was announced and immediately well received. It used the new ComStar wheels with triple disc brakes and the all-new 901cc model offered 23bhp over the original CB750. The CB900F Super Sports arrived in the UK the following year at a retail price of £2,099.00. The example on display in our museum is a 1982 USA model.

Launched alongside the Super Sports CB900F was a more sedate CB750K. Featuring a similar DOHC 16-valve 4-cylinder engine as the 900, the CB750K was aimed at the more relaxed touring rider. It had higher handlebars and more forward seated riding position, this model went on sale in the UK at £1,780.00 during the following year. The example in our display is a 1979 USA Limited Edition model, for which only 5,000 units were produced

▲ The stunning 6-cylinder DOHC 24-valve Honda CBX.

to celebrate Honda's 10th anniversary of the 750.

Honda's mid-range models had seen a number of changes over recent years but nothing quite so dramatic as the new Super Dreams launched this year. Continuing to feature the same engine as per the CB250/400T, the new CB250N and CB400N Super Dreams were treated to Honda's new Euro-Styling as featured on the CB900F. The 250 version in particular became a big hit in the UK appealing to the current 'L-Plate' rider laws. The CB250N went on sale this year at £799.00 and the CB400N at £949.00. The CB250N example on display is a 1980 UK type.

The previous CB400T Dream was re-launched with an automatic gearbox, similar to the 1976 CB750A, as the CB400AT Hondamatic. The system featured a two-speed semi-automatic transmission with a torque converter and two forward gears (high and low), manually selected by the rider. A parking brake replaced the clutch lever. The CB400AT Hondamatic went on sale in the UK at a retail price of £999.00. The example on display in our collection is a 1979 UK model.

Finally for this year, the Benly CM125T was announced. A custom design bike featuring the twin-cylinder engine of the Benly. Initially for the Asian markets, the model did reach our UK shores in 1982 as the CM125C, when the 'L-Plate' regulations were changed to 125cc.

81

1979

The extensive Honda R&D Tochigi Proving Grounds featuring a banked oval test course was completed. It was the most comprehensive private testing facility known at the time.

Soichiro Honda received an honorary Doctor of Literature degree from Ohio State University. Honda of America Manufacturing was established in Ohio and immediately started planning for production of the new GL1100 Gold Wing for the following year. Total production of Gold Wings switched from Japan to Ohio, where it remained for many successful years before reverting to Japan in 2010. Also, Honda Manufacturing was established in Nigeria.

Honda R&D had been extremely busy over the past year and a number of new models were announced and launched this year.

The Honda CB650 would be the last new model in a long line of successful SOHC fours. Based on the CB500/550 with the wet-sump engine the 650 provided great performance with light weight and regular, not sporty, riding position. With ComStar wheels, 4 into 2 exhaust and twin disc brakes up front, it was a great sensible all-rounder, that is often forgotten today. The CB650-Z went on sale in the UK at a retail price of £1,549.00.

The CX500 was given a custom makeover mainly for the American market where it went on sale first. It also remained to be known as the GL400 and GL500 Custom in some markets. It made it here to the UK and across Europe as the CX500C Custom from 1981 where it sold alongside the standard CX500 model.

The MB50 and MT50 Sports Mopeds (known as MB5 and MT5 in some markets) were announced for sales starting early next year and they immediately became firm favorites amongst sixteen year olds in the UK. A simple offering of a road sports model and an off-road styled trail model, cleverly using the same 49cc 2-stroke engine and similar tubular frame, which provided affordable transport for the new moped riders. At just £389.00 for the MT50 and £399.00 for the MB50 with cable-operated disc brake, they remained in the UK range until 1983. The example on display in our collection is a 1982 MB5 model from the USA.

On the back of the now successful DOHC CB900F a 748cc version was announced. The CB750F arrived in the UK during 1980 at a retail price of £1,780.00, £300.00 less than its bigger brother. We have a 1982 CB900F and a 1980 CB750F, both from the USA, on display in the museum.

The XL500S big thumper on/off road single came to the UK this year. A Custom version of the 400 twin was launched for various markets, but not the UK, as the CM400T. A new 2-stroke moped in the NX50 Caren, also known as the Express in some markets, was launched which came to the UK in 1982. The 2-stroke single-cylinder MB100, also known as the H100, was launched as a low-cost commuter which went on sale in the UK the following year as the H100 at a retail price of £409.00, compared to the current 4-stroke CB100N at £467.00.

A new range of CR off-road only motocross models was introduced. This included the CR80R, the CR125R Red Bullet and the CR250R Red Rocket. All three models were destined for the UK market where motocross was, and still is, a popular sport.

At Honda UK, we staged a special launch of these new motocrossers to dealers and press, which was held at the Hilton Hotel in Park lane, London. The main focus was on the CR250R Red Rocket and we designed a dramatic unveiling of the new model with yours truly dressed in motocross gear. I was perched on top of the bike at the top of a long ramp behind a giant screen on top of a 4ft high stage in front of a 250 strong audience. The idea was that a front cinema projection film of the new CR250R in action would play on the screen in front of the audience, and behind unseen, I would be released to run down the ramp and burst out of the film! What could go wrong? With only one paper screen we didn't have the option of a rehearsal, so only one shot! As I burst through the screen I had the coloured spot lights, projector light and dry-ice blinding my vision and all I could think was that there wasn't much distance before I was off the stage and into the audience! So I applied the brakes hard, which sent the carpet tiles on top of the stage firing into the audience, but I stopped with a fraction of an inch to go! Needless to say I didn't remove my crash helmet!

Towards the end of the 1970s, a new low-price moped was required to attract new customers to motorised two-wheel riding, just as the C100 Super Cub did in the 1960s. The NC50 Express, previously launched as the NC50 Roadpal for the Japanese market in 1976, was a simple design and easy to ride moped for both male and female first time riders. It had gone on sale in the UK the previous year at just £189.00 in a range of bright colours.

For the UK, a special TV advertising campaign was produced and launched starring the super-model, Twiggy. In my role I attended the filming, which took place outside Teddington Railway Station in Surrey. Twiggy arrived but hadn't been briefed that she was required to actually ride the Express. Not having ridden a motorised bike before, some instruction was urgently required! Well someone had to do it, so I spent a few hours teaching her, but of course it could have been done in 20 minutes! She fell off a couple of times, called herself a silly cow and got back on! From this TV campaign the low-price Express proved popular with new riders of all ages, and over 10,000 units were sold!

◀ The Twiggy Express as it became known and as can be seen in the museum. The previous famous ad line of 'You meet the nicest people on a Honda' helped generate sales of over 10,000 units.

83

The 1980s - 1980

The new XL and XR on/off road range of models were further extended with the addition of the XL50S, XL80S and a revised XL250S. For Enduro riders, the previously launched XR185, XR250 and XR500 models were joined by the XR200. All very capable on and off the road with powerful 4-stroke single-cylinder motors with the special 4-valve cylinder heads on the 250 and 500cc models.

Also added was the CB650C Custom, CB125JX, CB750 Exclusive and the 2-stroke Tact moped, later to be launched as the Melody in the UK. Further models included the CM250T, CD90A and the CB250RS, which became a UK favorite.

The CB250RS single cylinder sports bike was lightweight and offered nimble handling. Its slim design was a major feature. It featured a 4-valve cylinder head with twin exhaust ports to a 2 into 2 exhaust system giving the impression of a twin-cylinder engine! The CB250RS went on sale in the UK at a retail price of £759.00, a saving of £140.00 on the twin-cylinder CB250N Super Dream.

The off-road three-wheel all terrain bikes finally arrived in the UK with the launch of the ATC110, an enlarged engine model of the ATC90 as sold in the USA. An unusual vehicle to the UK market, the ATC110 sold for £580.00 and was purchased for leisure and agricultural use. Farmers could see the benefit of use in rounding up livestock and for reaching remote parts of their land.

I recall from this year, introducing three Japanese engineers from Honda R&D to a Yorkshire farm to further understand the use of the ATC in an agricultural environment. The engineers had never stepped outside of Japan before so the cultural shock of the farmer's

◄ The popular CB250RS – a single cylinder with a 4-valve cylinder head and twin port exhausts.

Britain's Graham Noyce ▶ competing in the 1980 World 500cc Motocross Championship having already won the 1979 series.

▼ Freddie Spencer aboard the 1980 Honda RCB1000 World Endurance bike.

84

wife serving them a typical English breakfast in their farmhouse kitchen was a sight to see! Full of normal Japanese courtesy, the engineers enjoyed the new experience and were surprised and pleased to see how their ATC creation was used to great effect on the farmland. In years to come, a range of ATC's were developed for more practical use, rather than leisure, so I think they certainly grasped the market potential.

Mainly for the USA market, the 6-cylinder CBX had only slight revisions for the 1980 model. These included a small lockable storage box in the rear seat cowling, reversed ComStar wheels finished in black and air-adjustable front forks. Known as the CBX-A or CBX'80, this model didn't make it to the UK formally, but several have been privately imported over time. The museum has a USA model example on display alongside the original CBX-Z model.

In racing, Honda won its first World Championship in the Motorcycle World Endurance Championship Series and in addition won the 500cc Motocross World Championship for the second year in a row.

Back in November 1977, Honda had declared its return to Grand Prix racing with a goal to enter the 500cc class in 1979. GP racing was dominated with 2-stroke machinery at this time. However, it was still Honda's wish and determination to decline the easy route, stay away from 2-stroke technology and fight the competition with 4-stroke engines. Under the project name of NR (New Racing), a 100 strong team of engineers had been assembled to undertake one of the toughest engineering challenges yet.

It had been determined that a conventional 4-stroke, 4-cylinder engine could not out perform its 2-stroke, 4-cylinder rivals. A Honda challenger had to be something different and special from the norm. Even to be level with the 2-stroke racers at the time, a 4-stroke would require twice as many cylinders and to rev to a minimum of 20,000 rpm! In addition, increased 4-stroke power would require more valves for reaching critical fuel-air mixture and conventional circular pistons could only accommodate a maximum of five valves.

By early 1979, after extensive experiments, the team had created the NR500 racer. It had an engine design that featured unique oval pistons in a V4 configuration allowing 8-valves per piston supported by two con-rods each, effectively a V8! During bench tests however, the engine produced just 90bhp, making it clear that they had much work ahead to reach their target of 130bhp. Problem after problem challenged the team and whilst the NR500 competed through this year and next, it went on to provide the basis for the stunning NR road bike of 1991. Honda finally had to give up the fight and revert to 2-stroke power to win the GP Championships of the 1980s.

Finally in this year, Montesa Honda S.A., a joint venture for the production of motorcycles, was established in Spain.

85

1981

▲ The VF750S – the first of many V4 models from Honda.

▲ The CX500 Turbo was the world's first mass-produced turbocharged motorcycle. It was upgraded to the CX650 Turbo in 1983 and this example is on display in our collection.

New models with new technology were in abundance this year with the announcement and launch of the CX500 Turbo; the V4 VF750S, GL1100 Gold Wing Aspencade; the CL250S Silk Road; the fold-away Motocompo; the XL250R and XL500R; the CB1100R; the CBX400F; the revised CBX 6-ylinder; a new range of ATC models and the 3-wheel NV50 Stream moped.

By now Honda had developed a habit of surprising the motorcycle world, and this year was no exception with the launch of the CX500 Turbo, the world's first mass-produced turbo powered motorcycle. The engine was based on the CX500 water-cooled V-twin but now fitted with a turbocharger that increased the power of the 500 to that of 1000cc models of the time. Also featuring fuel injection and on-board computer controlled systems, the CX500 Turbo was one of the most technically advanced motorcycles of the new decade.

Described as a 'technological feat in a beautifully crafted package', the new model featured a three quarter touring fairing, triple disc brakes, ComStar wheels and TRAC (Torque Reactive Anti-dive Control) front suspension. Also, new for this year, Honda's Pro-Link single shock absorber rear suspension system. The CX500 Turbo went on sale the following year in most markets and for the UK it had a retail price of £3,350.00. For 1983 the model was increased to 673cc. On display in our museum is a 1983 USA specification CX650 Turbo.

This was the time of a sales war between Honda and Yamaha. Honda's ammunition was launching new model after new model, with as much new technology as possible to show the world its supremacy in motorcycle creative design, technology and production.

Therefore, not content with just the new CX500 Turbo, Honda also announced an all-new V4 powered sports bike, the VF750S Sports also known as the V45 Sabre in the USA. The world's first liquid-cooled DOHC 90-degree 16-valve V4 engine, combined the high-revving power of an in-line 4-cylinder with the narrow width of a V-twin. The 90-degree angle of the cylinders also gave the engine perfect primary balance to eliminate vibration. The VF750S was a semi sports-tourer with raised handlebars, shaft-drive, new cast wheels and the new Pro-Link rear suspension system.

The VF750S Sports reached UK shores during 1982 with a retail price of £2,495.00. The new model signaled the start of many V4 models to follow. Whilst this one was shaft-driven, future examples would also feature chain-drive.

To add to the current ATC110, Honda UK introduced the ATC70 and ATC185S to the UK market. Honda R&D in Japan announced a 200cc version in the ATC200 for selected

▲ The NV50 Stream three-wheel moped as sold in the UK from 1982.

markets and the stunning ATC250R racer, which featured the same 2-stroke motor from the CR250R motocrosser. The ATC250R came to the UK in the following year.

The CBX 6-cylinder took a radical change in style and became a tourer. A touring fairing, panniers and the new pro-link rear suspension were added and the engine was finished in black. This CBX-B model (or CBX'81 as known in the USA) went on sale in the UK during the following year at £3,395.00.

A memorable super sports model announced was the race developed CB1100R. The limited-edition hand-made model, was based on the DOHC 16-valve air-cooled CB900F and the RS1000 factory endurance racer and only produced in limited numbers sufficient to meet race homologation. The 'R' was made famous in the UK by top British racers Ron Haslam and Joey Dunlop. The first model, the R-B, arrived in 1981 with a half-fairing and single seat (1050 units) followed by full-fairing dual-seat models of R-C in 1982 and R-D in 1983 (1500 units each), all very capable production racers or fast road bikes.

Also announced was the CBX400F, featuring a new DOHC 16-valve 4-cylinder air-cooled engine and Honda's new inboard disc-brake system, another first! This 400 was initially launched primarily for the Japanese market, but for the following year a 572cc version was introduced as the CBX550F and CBX550F2 with the latter featuring a half-fairing. Both of these models became available in the UK at £1,720.00 and £1,870.00.

For the Japanese market only, Honda R&D developed the NCZ50 Motocompo, a folding scooter to fit in the boot of a car. With a 49cc 2-stroke engine the strange looking compact commuter bike weighed only 45kg ready to ride! Not sold formally in the UK, a small number were privately imported. It was advertised on Japanese TV starring British music group, 'Madness'.

Also initially for the Japanese market, the unique 3-wheel tilting NV50 Stream was launched to generate new interest in motorised bike riding. The first of Honda's tilting three-wheelers, the Stream is a scooter-like single occupant vehicle with an automatic transmission and a 'one push' parking brake. It has a small-hinged rear pod containing the 49 cc 2-stroke engine and two drive wheels powered through a limited slip differential. The Stream was styled and priced as a luxury personal scooter. The NV50 Stream went on sale in the UK in 1983 at a retail price of £695.00.

Finally of note for this year was the launch of the CL250S Silk Road (also known as the CT250S). This was Honda's attempt at a 'trekking' motorcycle, marketed between its mechanical siblings, the CB250RS road bike and the XL250 dirt bike. It featured slightly more ground clearance than the CB250RS, and an upswept and close-fitted chrome exhaust that is kept clear of both debris and luggage. Its 6-speed transmission is geared as a regular 5-speed plus one extra-low gear. The CL250S was sold in the UK from 1982 at a retail price of £945.00.

1982

Far left:
The sports VF750F V4 known as the V45 Interceptor in the USA. This is a 1983 example from our collection.

The CX500E Euro Sport that later became the CX650E Euro Sport was also available as the CX400E in Japan.

Honda continued its policy of working to build motorcycles in various countries across the globe and signed technical collaboration agreements with Jialing Machine Factory of China, Peugeot Cycles of France and Montesa Motorcycles of Spain.

After its unsuccessful, but bold and inspiring attempt to win World Championship GP racing with 4-stroke machines, Honda Racing adopted 2-stroke technology for this year's World GP 500cc Championship. Immediately successful, they won the Belgian and San Marino GP's with rider Freddie Spencer on the all-new V3-cylinder NS500. Honda race wins are also achieved this year in the IOM TT F1, the Paris-Dakar Rally and the overall Trials World Championship.

A new super-economy Super Cub 50 was launched with a new engine that in controlled tests achieved a staggering 150km (93 miles) per litre, over 350mpg! The new Super Cub also featured updated styling with redesigned body panels and square lights. This new version came to the UK in 1984 as the C50-E Super Cub alongside its bigger brothers, the C70-E Cub and the C90-E Cub. By this time, the legal restriction on using the Cub name, held by Triumph Motorcycles as one of their Trade Marks, had expired.

Other new models included the VF750F, CX500E Euro Sport, the VT250F and the FT500. Also, the Spacy and Lead 125 scooters, the NU50 Runaway moped and the Gyro X.

Having launched the shaft-drive V4 VF750S the previous year, the new VF750F, using a chain-drive version of the same engine, was introduced as a more sporty option. Featuring an exposed box-section steel frame, headlamp fairing and under-cowling the new V4 came to the UK during 1983 at a retail price of £2,575.00, just £50.00 more than its shaft-drive cousin. In the USA this model was known as the V45 Interceptor.

Influenced by the styling of the CX500 Turbo, the CX500E Euro Sport was created and launched and went on sale in the UK at exactly half the price of its Turbo stable-mate. Like the Turbo, this model was also re-launched a year later as the CX650E Euro Sport.

Using similar technology from its new V4, a V-twin VT250F went on sale in Japan and came to the UK in 1983 at £1,395.00, £400.00 higher than the CB250RS of the time.

In recognition of the popular motorcycle sport of flat-track racing in North America, and named after the famous Californian venue, Honda created and launched the FT500 Ascot. A 498cc 4-valve, twin exhaust port big thumper single-cylinder engine as used in the XL/XR500 models and as used later in the 1984 XBR500/GB500 model. The FT500 was also sold in the UK from this year at a retail price of £1,350.00. On display in the museum is a 1982 FT500 Ascot example.

In Japan, Honda launched the first 4-stroke scooter of the 1980s. High fuel economy, quiet operation and reliability combined with great performance made the 49cc Honda Spacy such a big hit that R&D developed larger 125cc and 250cc versions the following year. The CH125 Spacy came to the UK by mid 1983 at £945.00 and the CH250 Spacy by 1985 at £1,399.00. In North America the Spacy models were known as the Elite model range with 80cc, 125cc, 150cc and 250cc offerings.

To run alongside the luxury 4-stroke Spacy models, Honda also launched a new range of 2-stroke scooters starting with 80cc and 125cc models, named 'Lead' in Europe and 'Aero' in the USA. The NH80MD Lead sold in the UK for £625.00 and the NH125MD Lead from next year at £775.00.

The NU50 Runaway moped was designed for male riders and used the 2-stroke engine from the Roadpal/Express. It featured a pressed steel frame and distinctive styling. Sold on both the domestic and overseas markets, it was nicknamed the Urban Express in some markets. It was on sale in the UK for just two years starting this year at just £365.00.

The Gyro-X was the second model in the tilting 3-wheeler series. Fitted with a non-slip differential mechanism and low-pressure wide-width semi off-road tyres, it enabled traveling on rough and snowbound roads. This model was mainly on sale

▲ The FT500 Ascot as sold in North America from this year and as on display in our collection.

in Japan and wasn't exported to Europe.

By now I had travelled to Japan on numerous occasions as the UK representative for product development. I had been privileged to visit R&D on many occasions to ride prototypes on their Tochigi banked oval test course and see new ideas being created and tested in the engine test-bays etc. I clearly recall always being so amazed at how many engine designs were produced as prototypes, run on test-bays and then either shelved or moved on to the next new model! It would, I'm sure, still be impressive today, some 35 years later!

1983

▲ The MVX250F alongside its racing relation, the NS500.

In this year the UK Honda range of models reached it's highest. From the 1983 Honda retail price list, some 82 different models were on sale compared to 57 models today! The dealer network also reached its highest, exceeding 600 dealer outlets across the UK compared to less than 100 today, 2017.

New models are again in abundance. Honda R&D and factory staff must have been consistently working overtime to produce such a large range. As the Honda vs. Yamaha sales war continued, the ongoing launch of new models remained Honda's key ammunition.

In celebration of their newfound World Grand Prix racing success, a production road version of the V3-cylinder liquid-cooled 2-stroke racer was announced with the MVX250F. Initially for the Japanese market only, the all-new model featured technology from the works NS500 racer, including a special crankshaft that balanced the left/right and center cylinders and flat-slide carburettors were also employed. Not on sale in the UK, the MVX250F did provide the basis of the NS400R V3-cylinder that was later launched and came to the UK in 1985 at £2,899.00.

The Gold Wing, now produced in Ohio, USA, was given a larger engine in the GL1200. The CB900F was joined by a bigger brother in the CB1100F which didn't unfortunately make it to the UK, being mainly sold in North America. The CX500T Turbo was also increased in engine size to the new CX650 Turbo. We have on display in the museum a 1985 GL1200 Gold Wing Limited Edition model, a 1983 CB1100F and a very early production 1983 CX650 Turbo with a frame number of just 69! All three models are from the US.

Success in the annual Paris-Dakar Rally provided the inspiration for the new XLV750R. It quickly became a popular model across Europe, especially France, but was not imported to the UK. The giant V-twin featured an engine that was derived from Honda's new liquid-cooled V-twin but converted to air-cooled. The rectangular steel tube frame contributed to the bike's distinctive look.

Two other new V-twins arrived this year starting with the VT250F, which had been announced for the Japanese market the previous year, but now available in other markets including the UK. A sophisticated 248cc sports bike, featuring a first in its class, liquid-cooled 90° V-twin engine. Characterised by its lightweight design, 16-inch front wheel, and inboard front disc brake, red frame, pro-link rear suspension and a headlamp fairing, the VT250F went on sale in the UK at £1,395.00.

Aimed at replacing the now dated CX500, the VT500E Euro Sport was the other new V-twin for this year (also known as the NV400 in some markets). It became available across Europe including the UK at a retail price of £1,695.00, £75.00 less than the outgoing CX500E Euro Sport, although stocks of the latter allowed availability for the remainder of this year. An in-line engine now powered the new model; a 3-valve per cylinder OHC liquid cooled 52 degree V-twin. It retained its shaft-drive appeal but lost its pro-link rear suspension to conventional dual dampers and also lost its twin front disc brake to the complex inboard ventilated single disc with a dual piston caliper. Despite its 30kg weight reduction, the new model sold in far less quantities to the more favoured CX500 models. The VT

▲ The V-four VF1000R with gear-driven camshafts, as on display in the collection.

engine however continued to make progress and featured in a number of new custom and sports models. These included the VT500C Shadow, the VT500T Ascot and later the VT600 and VT700 Shadows, all primarily for the US markets.

The in-line 4-cylinder model range was the subject of a further new variant with the announcement of the CBX750F. Retaining the traditional air-cooled transverse engine layout, the new model featured DOHC's and hydraulically operated maintenance free tappets, a new feature. Primarily sold in Europe, South Africa and Australia, the CBX750F featured as a similar model in the USA from 1984 as the CB700SC Nighthawk S. Although not related to the CBX 6-cylinder this model did adopt similar technology in mounting the alternator and ignition system behind the engine cylinders to reduce overall engine width. The CBX750F went on sale in the UK during 1984 at a retail price of £2,880.00.

The V4 range had another new addition this year with the announcement of the VF1000R, which arrived in the UK in the Spring of the following year. A heavyweight model, the bike's liquid-cooled V4 engine, which produced 125bhp @ 10,000rpm, was derived from the VF750F, but now featured gear-driven camshafts as opposed to the 750's chain-drive. Its stunning looks made for a dream superbike and even at 549lbs (249kg), the 'R' claimed at the time, the title of 'fastest production motorcycle in the world' with a top speed of 150mph. Its launch retail price in the UK in 1984 was £5,250.00. On display in our museum is a 1985 USA model example.

A smaller version of the CH250 was introduced as a CH125 Spacy for sale at £945.00. The ATC range was expanded with the announcement of the ATC200X utility trike. The Trial range was joined by the TLR200, which became a firm favorite amongst off-road competition riders and the TLM50 for young aspiring contenders. In addition, Honda UK started importing this year the Montesa-Honda's of MH200 and MH349 Trials bikes.

What was to become a famous long-standing model designation, the CBR first appeared with the announcement of the CBR400F, a Japanese domestic market small-capacity sports motorcycle. This 4-cylinder 16-valve, naked DOHC powered sports bike, featured air-cooling and a steel frame. It started the CBR sport model series as we know it today. It would be another four years before the CBR600F and CBR1000F were announced as 1988 models for the UK market.

Trail riding and 2-stroke single-cylinder engines provided the influence for the new MTX range of 50, 80, 125 and 200cc models as well as the road-sports MBX models of 50, 80 and 125cc with both styles of models utilising the same base engines. Both the MTX50 and MBX50 replaced the earlier MT/MB50 models. In addition, the GB250 Clubman nostalgic road bike was launched for the Japanese market.

Honda won the 500cc World Grand Prix Championship this year with Fast Freddie Spencer on the NS500. Honda returned to Formula One car racing with a new 1500cc twin-turbocharged engine, making its world debut at Silverstone.

1984

In the March of this year, I had the pleasure of inviting and accompanying around eight UK motorcycle journalists to a special test of new models in South Africa. The products on test were the new VF1000R, VF1000F, VF500F and the CBX750F. The schedule allowed for the spectating of the GP at the Kyalami circuit, testing of the new models on the same circuit the following day followed by a two-day ride out and back to the Kruger National Park. The main road to our destination was simply stunning, a new motorway winding through the hills with very little traffic! Little observation was paid to the local speed limits as we made our way to our over night stop in the game reserve. The return trip provided some excitement when one of the journalists was caught in a police speed trap at just over 150mph on the VFR1000R! Having to return to vouch for him at a remote police station, I found the rider being photographed and hailed a hero by the officers for holding a new top speed record that broke the previous by some considerable margin!

New models continued to be the order of the year and of course Honda R&D hadn't disappointed with many all-new designs. By this year the full range of Honda motorcycles featured power plants of all shapes and sizes. V2, V3 and V4 as well as singles, in-line twins, in-line fours, and flat-fours. Some with 2-sroke power but the majority with 4-sroke and in all sizes from 25cc to 1200cc. Over 100 different models were now in production for various worldwide market tastes and demands.

In May, Honda introduced two further new high-performance lightweight sport bikes, the NS250F and the NS250R. Both filled with bold style and a newly designed liquid-cooled 2-stroke 90-degree V-twin engine. Bristling with new materials and new technology the NS250 was developed using the latest sophisticated technologies gained from Honda's contender in the 1983 Moto GP World Championship, the NS500. The NS250F was a naked model using a steel frame while the NS250R was equipped with a full fairing on an aluminum frame. Sadly, mainly due to high price and a declining

The 90 degree V-twin 2-stroke NS250R.

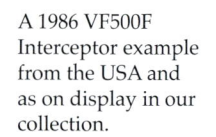

A 1986 VF500F Interceptor example from the USA and as on display in our collection.

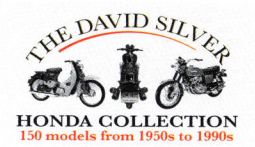

250cc market, these models were not formally imported to the UK.

For high performance models the focus remained on V4 technology. The new VF500F was announced to join the existing 400cc, 700cc, 750cc, 1000cc and 1100cc options. The VF500F, badged as 'Interceptor' for the USA market, was a mid-range sports motorcycle widely regarded as one of the finest handling motorcycles of the 1980s. Also available in a 'F2' version with a full rather than half-fairing, the VF500F2 arrived in the UK in the following year at a retail price of £2,550.00.

Towards the end of 1984, problems associated to the chain-driven camshaft system started to occur on the new V4 engines. Premature wear on camshaft lobes, rocker faces and camshaft bearing surfaces were being reported. Mainly affecting the 750cc models, as they were the biggest sellers, and with more claims in Europe, as the bikes were ridden harder there, the problems escalated into 1985. Honda's explanation drew attention to incorrect tappet adjustment and insufficient oiling to the camshafts at speed. New camshafts and rockers etc. were offered under warranty with replacement camshafts made from different material. Although the VF's could be made reliable again with the modifications the repairs were expensive which together with negative PR, cost Honda dearly. By now, it was becoming evident that the Honda vs. Yamaha sales war of the early 1980s, which Honda won in sales terms, was now taking its toll on Honda because of warranty claims. This was possibly due to rushed developments of new models and their shortened testing schedule in order to meet the strategy of more new models to win the war.

From the biggest petrol engine to the smallest, Honda challenged themselves with all sorts of engine design concepts. The new Honda People was one such example. Featuring a 2-stroke engine of just 24cc producing 0.7bhp in a pedal-cycle type frame with a total weight of just 24kg, the low cost moped provided economic daily transport to many Japanese commuters. The Honda people was primarily sold in Japan from 1984 at just ¥60,000 Yen (c. £189.00).

After its 25th year of production the Super Cub exceeded the 15 million units mark this year making it the most produced single automotive product in the world. Production continues to the present day in over 15 countries and has now exceeded 80 million units!

Honda signed a joint business agreement with Hero Cycles and its associates of India to establish Hero Honda Motors Limited for the production of motorcycles in India.

◀ The 25cc 2-stroke Honda People moped.

1985

Motorcyclists were treated to some authentic Honda GP bike action this year with the launch of the V3-cylinder 2-stroke NS400R, a new model derived from the works NS500. Often questioned why not the full 500cc, the 400 capacity model met the licensing laws of Japan where it was felt most of the demand would come from. I was also quietly informed that a full 500cc version would be far too powerful for general road/rider use!

In fact, Honda's largest ever 2-stroke motorcycle for the road was restricted to 59bhp for the Japanese market, but left unrestricted at 72bhp for the export market including the UK when it went on sale at £2,899.00. Finished in both tri-colour and Rothmans racing colours they were produced in low volume only. Finding one today in original top condition is quite rare and high prices can be expected.

American Honda celebrated the 10th Anniversary of the Gold Wing by offering the GL1200 Gold Wing Limited Edition model. Only c.5,000 units were produced for USA and Canada. This special model included fuel-injection, cruise control, self-leveling suspension, on-board computer and 4-speaker audio system. From July, production of all Gold Wing models switched from Japan to Honda's new USA Ohio plant. On display in our collection is an example of this rare model.

▲ The road-going GP derived NS400R with a V3 2-stroke power plant.

Having created and developed the almost perfect 498cc, 4-stroke, 4-valve single in the XL and XR range, the engine was transferred into a lightweight road sports model, the XBR500. Fitted with the RFVC (4-valve) cylinder head with dual exhaust ports that enabled the model to uniquely feature two separate chrome exhaust pipes and silencers. The XBR500, with retro café-racer styling, has both electric and kick-start arrangement which is linked to an automatic decompression valve to reduce compression during operation. The XBR was and still is a great lightweight fun bike with plenty of power through a 5-speed gearbox. It arrived in the UK with a retail price tag of £1,749.00. The example in our collection is a 1986 registered UK type.

Honda's line up of models also included a new range of competition motocross bikes from the CR50R to the CR500R. Also, a new pairing of Trials models in the TLR200 and TLR250 and continuing in the off-road department, a new duo of XR250R and XR600R Enduro models.

To run alongside the XL600R a tougher and larger XL600LM Paris Dakar with electric-start was made available to celebrate the annual endurance desert race that Honda finished 3rd and 4th in the last event. The XL600LM went on sale in the UK at £2,139.00.

Produced for just this year only, the VF1000F2 Bol d'Or was a full-fairing version of the previous VF1000F. Featuring a wind tunnel designed full sports-touring fairing, the F2 also

▲ The all-new V4 VFR750F with gear-driven camshafts.

adopted an extra radiator and a new seat design for long distance comfort for both rider and passenger. Engine modifications were carried over from the first 'F' models but the F2 retained the chain-drive camshaft layout. The VF1000F2 Bol d'Or came to the UK this year at £3,949.00, some £1,500.00 less than the VF1000R.

Following the previously mentioned high level of warranty claims on V4 models with problems relating to the chain-drive camshafts, Honda R&D had serious reservations about continuing with the V4 engine. They proposed a new range of in-line 4-cylinder engine models. However, at Honda UK and together with my counterpart at Honda Germany, we persuaded R&D to stay with the V4 concept and the result was the launch of an all-new model! This was probably the most important new model for many years. The all-new VFR750F was launched to the media in September at the Jerez circuit in Spain.

The VFR750F now featured a revised engine with gear-driven camshafts, which addressed the previous camshaft issues. It also featured an all aluminium twin-spar lightweight frame, a 16-inch front wheel, wide profile tyres, 6-speed transmission and a full fairing. The new generation V4 was finished in a pearl-white option that looked stunning at its launch, where I was pleased to be in attendance, and the new 750 was met with praise by all the European journalists. This was Honda engineering at its best as the new model quickly dispelled the poor VF image and set new standards for the future. The new VFR did not share one part from the previous VF750F and with more power, less weight, exquisite looks and stunning ride, it was 110% better!

▲ The complex heart of the VFR750F.

The VFR750F arrived in the UK the following year at c. £3,500.00. At the Easter Transatlantic races, Ron Haslam took a stock VFR, straight from the Honda press fleet, to a podium finish at the rain soaked Donington Park race. For the USA, the VFR retained the Interceptor naming. As was traditional by now, they also had a 700cc version as a 'Tariff Beater' against the local tax laws for over 700cc motorcycle imports. This was introduced to protect the then troubled Harley Davidson Motor Company.

This first generation VFR750F sold successfully in most markets and provided the basis of a number of generations of VFR's to follow. Including of course, the rare limited edition VFR750R (RC30) that took honors in many road races around the world, including the IOM TT. Still based on the original platform the VFR continues today, some 30 years later, in the VFR800. It remains my personal favourite with one taking pride of place in my garage!

1986

The all-new VFR750F went on sale and while initial reaction and order books were exceeding expectations, Honda's Sales and R&D teams were unsure of its future at this stage. They had determined that they also needed a more cost-effective range of sports models.

Behind the scenes, and having almost run in parallel with the VFR development as a safe guard, a new range of DOHC liquid-cooled in-line fours were in progress. Announced later this year, the new CBR range had started with the new CBR250F that featured gear-driven camshafts and an alloy frame. It went on sale in Japan only during this year with a half-fairing and was later joined by a CBR250R full-fairing version. A 4-cylinder, liquid cooled, DOHC 4-stroke in an aluminium frame with race style fairing! What more could the 250 riders of Japan ask for? Needless to say it became a big hit even at the fairly high price!

For export markets, the new CBR family was later joined by the CBR600F, sold as a 500 in some markets and also as a 400 in others, and the CBR1000F. To keep costs acceptable, these models featured a chain-drive camshaft arrangement and a steel frame hidden behind the enclosure of the full bodywork. Quickly accepted as reliable, high-performance superbikes the CBR600F produced 85bhp @ 11,000rpm and the CBR1000F, 135bhp @ 9,250rpm. The 600 was soon adopted as a

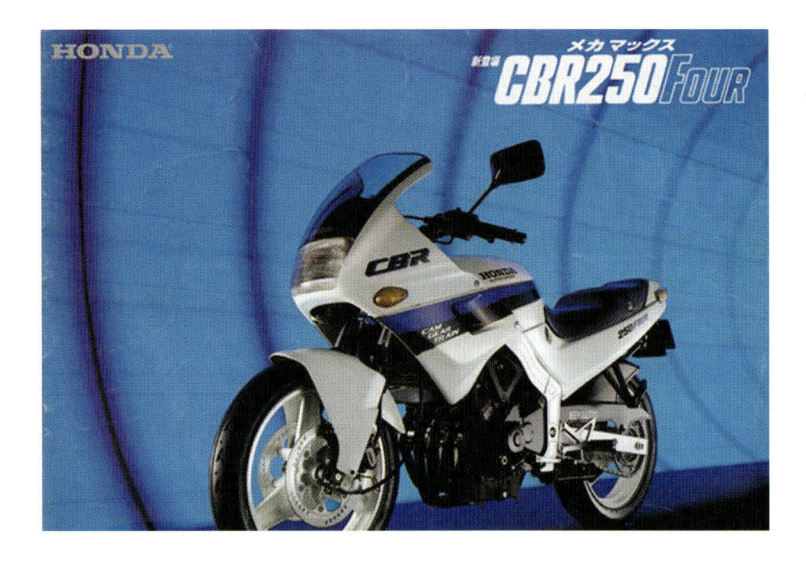

◄ The DOHC 4-cylinder liquid-cooled CBR250 Four for the Japan market only!

◄ A USA example of the DOHC 4-cylinder liquid-cooled CBR1000F Hurricane as on display in our collection.

production racer by many budding young riders with one winning the IOM TT production race for me. I had Scottish rider Brian Morrison on board, under my Bike Studio dealership banner. This was the Honda dealership I started after leaving Honda UK.

The 600 model was also selected for the new for 1987 Honda CBR600 Challenge. A one make, one model, same specification race series that ran across UK circuits for two years. With no individual power advantage, the racing was very close with a number of individual race winners.

On sale the following year in most markets the CBR600F arrived in the UK at £3,299.00 and the CBR1000F at £4,399.00. The CBR750F also joined the range later, for selected markets, totaling six different size models in the CBR series. Highly successful, these models remained in the Honda range for a number of years until the launch of its next generation, the CBR900RR Fireblade in 1992.

Also, the VF1000R was upgraded with Rothmans race colours, the main sponsor of the Honda race teams. The CB350S was introduced to the UK market at £1,899.00. It featured an engine based on the earlier Dream twin and an unusual red finished tubular steel cradle frame.

Since 1984, a special team of Honda R&D engineers had been formed to develop an all-conquering bike for the annual Paris Dakar Rally with the target of winning the 1986 event. After considerable development the new model named NXR was completed and went on to take 1st, 2nd, 3rd, 5th and 6th places in this years Paris-Dakar Rally motorcycle competition. The special engine with unique event specific characteristics, was a V-twin based on the RS750D engine which Honda developed for dirt-track racing in the USA.

It was this year that David Silver established his Honda parts business under the trading name of David Silver Spares. Starting in North London, David later moved his business to Leiston, Suffolk. (See page 10 for further background details).

▲ The CBR600F as used in the 1987/88 rounds of the Honda CBR600 Challenge one-make/model race series.

Honda wins the Paris-Dakar with the V-twin NXR Africa Twin and Cyril Neveu of France. ▶

97

1987

Honda RC30 – today, probably the most iconic road-going sports/racing motorcycle ever produced. The VFR750R, factory product code RC30, was first announced this year and initially went on sale exclusively on the Japanese market. The most famous of the VFR series, it is a fully faired road-legal racing motorcycle, produced in limited numbers for World Superbike racing homologation purposes by Honda R&D and HRC (Honda Racing Corporation).

Starting with a restricted power version (77bhp @ 9,500rpm) on sale in Japan, the full power (112bhp @ 11,000rpm) RC30 arrived in the UK during 1988. It had a retail price of £8,699.00, almost twice the price of the standard VFR750F! However, the RC30 was very special and today they can demand prices north of £20,000.00. In fact, one low mileage showroom example sold for over $40,000 in a 2015 USA Bonhams Auction.

◀ The Honda VFR750R aka the RC30.

The 748cc, 16-valve gear-driven DOHC liquid-cooled 90° V4 contained race-inspired components from the works RVF endurance racer of the time. Engine components included such items as titanium connecting rods to reduce reciprocating weight, special pistons and gear driven camshafts.

Whilst similar in external appearance, the engine of the RC30 shared no internal parts with the road going RC24 (VFR750F). The engine firing configuration was very different from the RC24 from which it was derived, with a 360° 'big-bang' crank arrangement instead of the smoother 180°. This feature produced a very broad spread of power and, when coupled to the close ratio gearbox made the RC30 untouchable in terms of drivability. Slowing down was made easier with a slipper clutch, and impressive braking capability for the period. It redlined at 12,500rpm, compared to 11,000rpm on the RC24 and weighed 180 kg (400lb) dry, some 36kg (76lbs) lighter than the RC24.

On the chassis side, the RC30's lightweight aluminium frame was fitted with front suspension made by Showa with wheel and brake pads that had quick-release mountings. The rear wheel carried a brake disc to the inside and a chain sprocket to the outside of a single-sided swingarm (originally patented by ELF of France).

The bike was also equipped with fully adjustable Showa rear suspension that gave superior ride and handling characteristics, especially as it was set-up for solo riding only. The engine and low storage position of the fuel in the fuel tank combined to give a low centre of gravity, which also aided its handling ability. Further statements of its hand-built quality were shown in a full stainless steel 4-2-1 exhaust system, alloy fuel tank and hand-laid fiberglass bodywork.

The popular Honda Vision mopeds already on sale in the UK, known as the Tact on the Japanese market, were further enhanced with the addition of a compartment under the seat to store a safety-helmet. This was now required to be worn by law in many countries. Known as the SA50 Vision Met-in (short for helmet in), the new model arrived in the UK in early 1989 at £839.00.

This would be the last year of production of the flat-four cylinder Gold Wing. A revised GL1200 Aspencade equipped model was produced and went on sale in the UK in limited numbers at £8,399.00.

Reverse gear and ABS braking on motorcycles were amongst new technologies being developed by Honda R&D at this time. The company announced the worlds first antilock braking system (ABS) for motorcycles during this year and then first featured ABS on a new European touring model being developed for 1990. Reverse gear however, was to come earlier on an all-new GL touring model for 1988.

Honda becomes the world's first company to produce 50 million motorcycles with cumulative production at Kumamoto Factory alone reaching the 10 million unit mark. In automobile engineering, Honda produces the world's first production 4-wheel steering system as featured on their Prelude sports car.

In motorcycle racing, Honda took 1st and 2nd places at this year's Paris Dakar Rally. In car racing Honda F1 engines took the first four places at the British Grand Prix held at Silverstone.

▼ The SA50 Vision Met-in with built in under seat safety-helmet storage.

1988

The all-new GL1500 Gold Wing went on sale this year. The 4th generation Gold Wing was now powered by a 6-cylinder engine featuring a reverse gear! At 794lbs (360kg) it needs it!

A specialist design team spent over three years developing the new model, which is one of the longest periods spent on one new model. The GL1500 had an ambitious design target, which was reached with full approval. Honda described prototype testing as involving sixty development stages, and building fifteen different test bikes. This included one made from a GL1200 frame coupled with the original prototype engine so that a 6-cylinder could be compared to the 4-cylinder head-on. New Gold Wing design goals were: smoothness and quietness coupled with enormous power and a high level of specification with sophistication.

The new GL1500 Gold Wing had made its debut at the 1987 Cologne Motorcycle Show, 13 years after the original GL1000 was first shown at the same event. The 1988 model brought the most changes seen to the Gold Wing series since its inception. Although the new 6-cylinder still used carburettors, there were just two large 36 mm CV Keihins supplying all cylinders, the first time any Gold Wing had less than one carb per cylinder.

The new Wing was enclosed entirely in plastic, giving it a seamless appearance. The seat height was at its lowest yet on a Gold Wing. The passenger back-rest, trunk and saddlebags were integrated with a central mechanism locking the lot. An on-board compressor adjusted rear suspension air pressure and the addition of the 'reverse gear' was actually a creative use of the electric starter motor linked to the transmission. Nearly every dimension of the bike had grown; a larger windshield, longer wheelbase, two more cylinders, more horsepower, more bodywork, more electronics, more accessories and more mass. The 4-speed plus overdrive GL1500 also featured a sophisticated electronic cruise control system. Options included a passenger audio control and rear speakers, CB radio, auxiliary lights and exterior trim.

◄ The 6-cylinder GL1500 Gold Wing featuring reverse gear and cruise control.

◄ The single-cylinder NX650 Dominator.

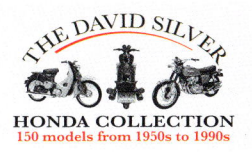

Although primarily designed for the North American markets, the GL1500 Gold Wing also arrived in Europe including the UK at a hefty £9,199.00. The Ohio USA built Wing was also exported to Japan!

The VT500E was replaced with one of the first of a new breed of naked-bikes. An all-new SOHC 3-valve narrow-angle V-twin was launched in several versions. For Europe, we had the NTV650 Revere with shaft-drive and steel frame and for America, they had the NT650 Hawk GT with chain-drive and aluminium frame. Japan had the NT400 and NT650 Bros (the naming derived from and in reference to the Brothers of 400cc and 650cc versions that were conceived at the same time) also with chain-drive and aluminium frame.

Regarded as a cousin to the Hawk GT, the Revere was available in the UK from this year at £3,299.00. It had an integrated steel frame, larger fuel tank, a lower and stubbier silencer, shaft drive, single-sided swinging arm with adjustable monoshock suspension and a centre stand.

All these NT based models shared the same engine and many other components and were blessed with fine handling and low maintenance (especially with the shaft drive) and little vibration due to the use of offset crank-pins. The NT was only the second Honda model to feature the Pro-Arm single-sided swingarm/rear suspension system, after the RC30.

The example on display in our collection is a 1988 USA NT650 Hawk GT model. Based around the same NT engine, the USA markets also enjoyed the launch this year of a new custom model in the VT600C Shadow. However, the frame and all cycle components were unique to this model.

The NX650 was a further new model that started a new range of NX models to replace the XL series. Enlarging the successful SOHC RFVC single-cylinder motor to probably its maximum, the 644cc power-plant provided substantial torque for this dual-purpose on-off road machine. The original Japanese market model featured both electric and kick starting but this was amended to electric only on all other models. The NX650 is unique in that it carries the engine oil in the frame. Its image started a cult following due to a fashion of owners converting them to 'Supermoto' style. The NX650 Dominator arrived in the UK at £2,999.00

In addition to the 650, there were also NX125, NX250, NX350, NX400 and NX500 variations for various world markets, the NX350 was produced in Honda's Brazil factory during the 1990s.

During this year, a private team of Honda Suzuka Factory employees designed and built a special Honda powered vehicle for the annual Shell Mileage Marathon in England. Today known as the Shell Eco-marathon, it set a new world record of 2,269km/litre, equivalent to 5,337mpg! The team went on to win again in 1989 and 1990.

◀ The V-twin NT650 Hawk GT, a 1988 USA specification model as on display in the museum.

1989

The RC30 inspired NC30.

No all-new models this year as many of the current popular models such as the Transalp 600V, NTV600 Revere, CBR600F and CBR1000F received technical updates and cosmetic changes. However, behind the scenes, Honda R&D were busy with a number of exciting new models for the following year.

Launched for the 1990-year sales in various countries, including the UK, was the third generation of the VFR400R, the best known of the series. Like its bigger brother the RC30, the VFR400R was also popularly known by its NC30 factory code. VFR400's had been on sale in Japan for a number of years by now, as the Japanese market was influenced by its licensing laws favoring up to 400cc models. For this third-generation model styling, changes produced a scaled down copy of the RC30, it's often difficult to tell them apart. This generation model also saw the introduction of the 360° crank firing engine, which is known as the 'big bang' engine.

Forever popular in Japan, the NC30s were available in a total of eight different colour schemes, produced with three different model year specifications from 1989 to 1992. Export models were made in two different colour schemes, and carried model year designations L and M for 1990 and 1991. Although expensive and therefore limited in sales numbers, the VFR400R NC30 is widely regarded as being a fine handling motorcycle. The engine has a very wide power-band for a 400cc engine, which made it a perfect beginners superbike. It has a cult following in Japan and the UK, where the NC30 is a popular track-day and racing bike.

For the growing UK learner market, the taste for 125cc bikes that looked like superbikes was a mission for the Honda designers. With Honda's factory in Italy now producing a number of models up to 125cc including the NS125F and NS125R that had been on sale in the UK, the factory produced and launched the new NSR125R that had the styling of a GP racer and the dimensions similar to that of a superbike! The NSR125R went on sale in the UK during the following year.

As the Gold Wing had grown in size, so had the price, so there was a demand for a lighter more affordable all-touring model. A special team of Honda designers and engineers had been busy for some time developing the all-new ST1100 Pan European that was announced this year for 1990 sales.

Featuring a 1,085cc longitudinal V4 DOHC engine, shaft-drive, full touring fairing and matching pannier set, the ST1100 Pan European was another all-new model surprise from Honda. It was not based on anything seen before as the designers started with a clean sheet to create a long-term popular touring model for the European customer.

Specification was high, as the ST1100 was created for long-distance touring with low-maintenance, comfort and safety.

It featured a huge fuel-tank providing a range in excess of 250 miles and decent sized panniers for holiday trips. The Pan-European was a hit from its original launch establishing a loyal following of owners, many of whom joined the soon to be launched, Honda organised annual Pan-European gatherings in various countries. From these early highly organised tours, Pan-European clubs have spawned throughout Europe. Today with the original ST1100 and now the ST1300, the interest and following is just as high with regular owner rallies across the continents.

On the back of their winning success in the annual Paris Dakar Rally, Honda announced in December the XRV750 Africa Twin, to go on sale in early 1990.

The XRV750 Africa Twin was a dual-sport model, based on the factory competition NXR-750, which won the Paris Dakar Rally four times in the late 1980s. It was built in tribute to the giant desert racers of the rally. The Africa Twin features a large, 742cc, 6-valve, 4-spark plug V-twin liquid-cooled power plant. It has twin headlights, a windscreen, and a long dual seat that stretches back from the tank to its luggage rack. An aluminium bash plate protects the bottom of the engine from flying rocks and impacts. With long travel suspension, twin disc brakes up front and single at the rear fitted to spoke wheels, the Africa Twin became an instant hit with European riders. Today, the latest generation example is one of Honda's best sellers.

Throughout the 1980s, Honda R&D had been developing its next generation of 4-stroke engines, primarily for cars. However, new technology was always developed at Honda with all products in mind.

As part of their new generation engine, which was being launched in the 1989 Honda Integra model car, the system of

▲ The Paris-Dakar developed XRV750 Africa Twin.

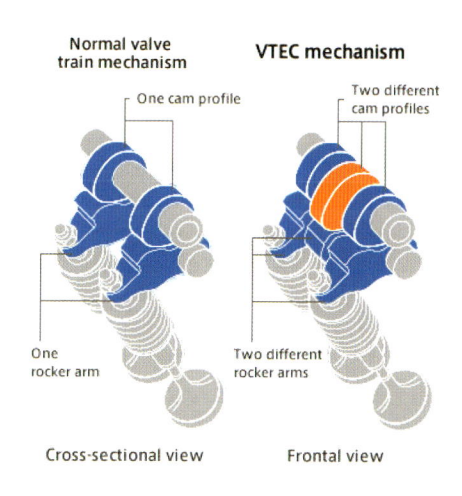

▲ Inside the VTEC system.

VTEC (Variable valve Timing and lift Electronic Control system) had been developed for more power and better fuel economy. It was innovative technology that surprised many automotive industry leaders with a new level of performance.

The VTEC system was the installation of a new set of cam followers and rocker arms for high-speed operation on the intake and exhaust sides that, along with the switching of cam hills according to engine speed, provided higher engine efficiency. VTEC was later adopted to motorcycle engines as seen on the VFR800, which is a prime example of Honda's ongoing combined all-product engineering development.

The 1990s - 1990

The VFR750R (RC30) finally went on sale in North America, three years after it was first sold in Japan. At twice the price of the VFR750F Interceptor, it soon found its investors hoping for long-term gain, due to its appeal and limited production numbers!

As a Honda dealer, I was invited by Honda UK to a remote hotel in Wales for a presentation and launch of the new ST1100 Pan European and more excitingly, to collect our dealer demonstrator and ride it home.

All UK dealers who had signed up to sell the new tourer were invited over several days to the same location where a large number of Pan European's in both deep red and silver were prepared and waiting for us. After a technical presentation, evening dinner and over night accommodation I set off the following morning for the ride back to North London accompanying some Honda dealer friends whose destination was a bit further to Essex. Initially, I recall not being too sure of the ST1100's ride! I actually stopped to check the tyres and to make sure all was well. It took some getting used to, as its ride characteristics were of course very different to that of the CBR1000F and the Gold Wing, Honda's other large capacity bikes of the time. After some 50 miles into the 150-mile journey I had finally adapted my ride accordingly and started to enjoy its pleasant smooth touring pace, comfort and many other ride qualities. The ST1100 Pan-European went on to become a great success.

By 2002 the second-generation model became the ST1300 and despite being a model developed for Europe, it was also introduced to America a year later.

After four years in the market, the popular VFR750F (factory product code RC24) was continuing to sell well

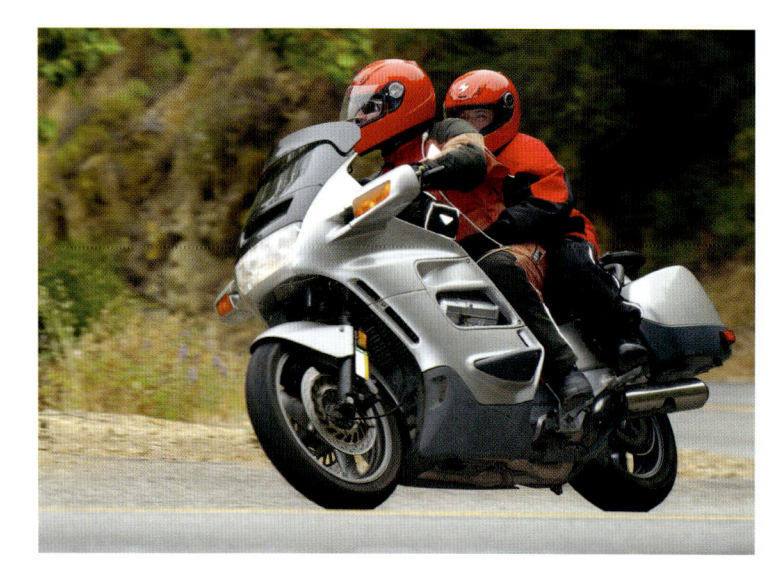

◀ The V4 ST1100 Pan European.

◀ A 1990 USA version of the VFR750F as on display in our museum.

and following some minor changes, such as 17-inch front wheel from the original 16-inch, it now received a major revamp.

The 1990 VFR750F (RC36) was redesigned with a new frame and bodywork, revised front forks, the new single-sided swingarm and wider wheels for modern tyre sizes, but it did gain a bit more weight! The new model was soon on sale, boosting Honda's big bike sales even further. A 1990 USA model of the VFR750F is on display in our collection.

This was the year that I sold one of these new VFR's to Pink Floyd drummer and classic car collector, Nick Mason. I recall delivering the bike to his central London office above his garage that contained an incredible collection of exotic Ferrari's and other collectable cars. But biking was clearly his passion! On that note it has been surprising over time to learn how many celebrities like to use motorcycles as with their helmet on, they can avoid the limelight!

For the top-end of the commuter market a new 244cc 4-stroke liquid-cooled single-cylinder powered scooter was launched in the UK as the CN250. This model had actually been on sale in the USA since 1986 where it was known as the Helix and in other parts of the world as the Fusion or Spazio. It had replaced the previous 4-stroke scooter that was known as the CH250 Spacy in Europe.

The new CN250 model lengthened the CH250 by some 14 inches (360 mm), it featured an integrated trunk in the rear of the scooter and a low seat height. The added length allowed a feet-forward riding position and a smoother ride than that of previous models. The top speed of the machine was limited to 75 mph (121 kph), but the drivetrain was of an under-stressed design allowing extended running at top speed. The CN250 was a very capable machine complete with trip odometer, fuel and temperature gauges, glove compartment and trunk. It remained so successful that it remained in production in Japan for over 20 years.

In racing, Honda won their first 2-stroke 125cc World Championship with the RS125R, and the factory V4 RVF750 won Honda's sixth consecutive victory in the popular Bol d'Or 24-hour race. Honda also took championship titles in the 500cc World Motocross and World Superbike.

▼ The RVF750 on its way to winning the 1990 Bol d'Or 24-hour race.

1991

Honda management in Japan began a major corporate reorganisation this year that included the separation of the Honda business into three distinct product lines: Motorcycles, Automobiles and Power Products.

Despite the previous NR500 and NR750 racing projects of the 1980s ending in failure, a considerable amount of knowledge had been gained from the use of new materials and from the bikes brave technology of using oval piston engines. As previously mentioned it was always Honda's belief that what was learnt on the track should be translated to the road and the V4 oval-piston technology was no exception.

Honda was in the final stages of developing a limited edition NR road bike for the 1992 sales year. The NR (RC40) was a very special and very expensive 748cc V4 road bike that was more a showcase of Honda's engineering ability, rather than a serious production bike. Only 300 units were produced worldwide and only around 30 of those came to the UK. I was very lucky to have gained one for my dealership in North London, which was bought by a young chap from Belfast who arrived with the required £38,000.00 in cash in two Tesco shopping bags! My chief mechanic was not keen to ride the valuable and rare exotica, as required in the pre-delivery inspection (PDI), in fear of scratching it, so yours truly had the honor! If I'm honest I couldn't feel the extra £25,000 in the ride compared to the RC30, in fact the RC30 felt just as good if not better, and there lies the real reasoning of the NR!

I suspect only a handful of the 300 NR's produced have exceeded 10,000 miles in total use. The real interest is in having one as asset value. Today, they only occasionally change hands at over £60,000.00 and some have even been advertised at over £100,000.00! For those who bought one at the original price, are enjoying the view of a beautifully engineered Honda that will only continue to grow in value.

Often referred to as the NR750, but formally named as just NR, the rare Honda was not only the most expensive production motorcycle of the time but the most complex and most difficult to build on the production line. In fact, they were all hand built by individual working groups. Regarded by some as too complex, beyond the reach of the regular motorcyclists, the exquisite looking, hi-tech superbike with extremely high levels of build finish, will remain in the history books for many years to come.

The Honda CBR900RR Fireblade was another development sensation from 1991. Scheduled for the 1992 sales year, the new in-line DOHC 4-cylinder was to become one of the most successful superbikes of all time. This was the first generation model with an 893cc engine

◀ The NR production road bike as sold in 1992 in the UK for £38,000.00.

◀ The CBR900RR Fireblade - still a favorite and a winner on road and track today and as on display in our museum.

with factory code SC28. Developed over a long period of time by Tadao Baba and his team, a name that would became synonymous with the Fireblade. The CBR900RR set a precedent for lightweight in the superbike class. At only 205kg, it was just 2kg heavier than Honda's own CBR600F2 but more importantly, 34kg lighter than the Yamaha FZR1000 of the time.

The launch of the Fireblade was Honda setting yet another new standard in superbike production. Throughout the 1990s and the 2000s, to have a Fireblade was to have one of the ultimate road bikes. Today, some 25 years later the Fireblade, now a CBR1000RR, continues to lead as the flagship in the Honda motorcycle range.

The final development and announcement of the NR and the CBR900RR was a fitting testament to the Old Man. In August, having developed his company from a small operation with just 12 staff to the world leader in motorcycle development and production, founder Soichiro Honda passed away at the grand age of 84 years.

This is currently the last year we cover in our Honda motorcycle museum display and the time that we also end our year-by-year history of Honda motorcycles. It's an appropriate time with the passing of the great engineer and visionary leader that was, Soichiro Honda.

▲ Soichiro Honda, 1906 to 1991.

Today and Tomorrow

2016 marked the 70th year since Soichiro Honda started his motorcycle business under the name Honda Technical Research Institute before renaming the company to Honda Motor Company Limited in 1948.

The company has come along way at a very impressive pace. It had its struggles but it's had many successes much of which must be contributed to Soichiro's philosophy to life.

Today, the company is a world leader in many automotive and engineering fields:

- Motorcycles
- Automobiles
- All-Terrain Vehicles
- Generators
- Agricultural Equipment
- Lawn Mowers
- Marine Engines
- Robotics
- Aircraft

Honda has also conquered motor sport in many disciplines since the 1960s.

- Isle of Man TT Races
- World Motorcycle GP
- World M/C Endurance
- World Trials
- Paris Dakar Rally
- World Motocross
- Formula 1
- IndyCar
- World Touring Cars

…and many other international motor sports.

Motor sport remains Honda's most important testing ground. From the 1960s Soichiro Honda firmly believed in this principle, quoting then…'Racing improves the breed'. If you look at the evolution over 50 years of the first superbike, the 1968 announced Honda CB750 Four developed from the factory GP race bikes, and todays CBR1000RR developed from the latest factory Moto GP bike, you can see how his principle has not only worked but continues today, some 50 years later.

Year	Model	Engine	Power
1968/69	Dream CB750 Four	SOHC 8-valve 4-cyl Air-cooled	68bhp @ 8,500rpm
2017/18	CBR1000RR Fireblade	DOHC 16-valve 4-cyl liquid-cooled	133bhp @ 12,250rpm

Honda has probably become the most successful automotive and engineering company of all time, and they haven't stopped!

Alternative fuel has been on Honda's agenda for many years. They were one of the first companies to develop and promote Hydrogen fuel cell for cars and are today fast developing electric motors for cars, motorcycles and other products in their range.

What many readers may not know is that Honda has long been associated with Robotics. They introduced ASIMO in the year 2000, a 4ft 3inch humanoid robot that today plays football and table tennis, dances, runs and talks in many languages, as well as helping those without full mobility. ASIMO is just one product example of Honda's wealth of knowledge in Robotics.

In 2015, another Soichiro Honda dream was realised, when full production and sales started of Honda's six seat corporate jet aircraft known as the HA-420 HondaJet. Designed in Japan and built in Honda's American aircraft factory and with a maximum speed of just under 500mph (804kph), this is the fastest and most expensive Honda product yet. At a cool $5million, you could add one to your Honda fleet!

Honda products are now manufactured and assembled in over 20 different countries and sold in over 150 countries. In November 2014, Honda reached a milestone in producing its 300 millionth motorcycle, which included over 80 million Super Cub models!

For tomorrow, who knows? With an extensive motorcycle model range to meet all demands and tastes and an R&D department that continues to push the envelope of technology and design we can only dream of the Honda products of the future, just as Soichiro Honda did back in 1946.

An example of todays Honda motorcycles - The 2016 NM4 Vultus demonstrating iconic styling with precession engineering.

Honda ASIMO with the HondaJet…and no, ASIMO wasn't the pilot!

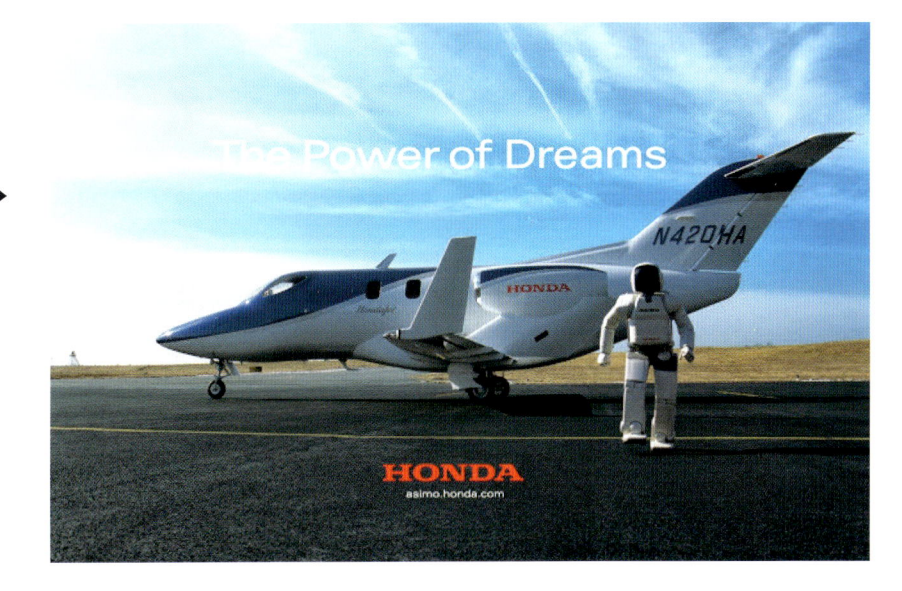

In Superbike, Honda hasn't lost its touch with the race-bred all-new CBR1000RR Fireblade SP for 2017/18. 50 years of amazing Superbike evolution from CB750 to CBR1000RR.

109

General specifications of the models on display in The David Silver Honda Collection:

Specifications:	Cub F-Type	Dream E-Type	Dream SA	Dream ME	Dream CS71	Dream CE71	Benly J-Type	Benly JB	Benly JC56
Model on display:	1952 Cub F-Type	1953 Dream E-Type	1956 Dream SA	1957 Dream ME	1960 Dream CS71	1959 Dream CE71	1954 Benly J-Type	1955 Benly JB	1956 Benly JC56
Year of introduction:	1952	1951	1955	1956	1958	1959	1953	1955	1955
Engine Type:	1-cyl, 2-cycle	1-cyl, OHV, 4-cycle	1-cyl, SOHC, 4-cycle	1-cyl, SOHC, 4-cycle	2-cyl, SOHC, 4-cycle	2-cyl, SOHC, 4-cycle	1-cyl, OHV, 4-cycle	1-cyl, OHV, 4-cycle	1-cyl, OHV, 4-cycle
Capacity:	50cc	146cc	246cc	246cc	247cc	247cc	89cc	125cc	125cc
Power:	1bhp @ 3,600rpm	5.5bhp @ 5,000rpm	10.5bhp @ 5,000rpm	14bhp @ 6,000rpm	20bhp @ 8,400rpm	18bhp @ 7,400rpm	3.8bhp @ 6,000rpm	4.5bhp @ 5,500rpm	7bhp @ 6,000rpm
Transmission:	Variable 1-speed	2-speed manual	4-speed manual	4-speed manual	4-speed manual	4-speed manual	3-speed manual	3-speed manual	4-speed manual
Frame:	-	Channel pressed steel	Pressed steel	Pressed steel	Pressed steel	Pressed steel	Pressed steel	Pressed steel	Pressed steel
Suspension, Front/Rear:	-	Telescopic/Plunger	Telescopic/S.Arm	Leading-link/S.Arm	Leading-link/S.Arm	Leading-link/S.Arm	Telescopic/S.Arm	Telescopic/S.Arm	Earles/S.Arm
Brakes, Front/Rear:	-	Drum/Drum	Drum/Drum	Drum/Drum	Drum/Drum	Drum/Drum	Drum/Drum	Drum/Drum	Drum/Drum
Wheels, Front/Rear:	-	19inch/19inch	19inch/19inch	18inch/18inch	16inch/16inch	16inch/16inch	24inch/24inch	24inch/24inch	19inch/19inch
Weight, Dry:	-	214lbs (97kg)	376lbs (171kg)	383lbs (174kg)	348lbs (158kg)	348lbs (158kg)	209lbs (95kg)	220lbs (100kg)	242lbs (110kg)
Top Speed:	22mph (35kph)	47mph (75kph)	62mph (100kph)	62mph (100kph)	84mph (135kph)	84mph (135kph)	40mph (65kph)	50mph (80kph)	50mph (80kph)
Intro Price: (Market'Yr)	¥25,000 (Japan'52)	N/A	N/A	¥169,000 (Japan'56)	¥182,000 (Japan'58)	N/A	¥99,000 (Japan'53)	N/A	N/A

Specifications:	Benly JC57	Benly C92	Benly CB92	Benly C95 Touring	Super Cub C100	C240 Port Cub	Little Honda P50	Little Honda PC50	PS50K
Model on display:	1957 Benly JC57	1963 Benly C92	1962 Benly CB92	1965 Benly C95	1961 C100	1962 C240	1967 P50	1969 PC50 (Japan)	1969 PS50K
Year of introduction:	1956	1959	1959	1958	1958	1962	1967	1969	1967
Engine Type:	1-cyl, OHV, 4-cycle	2-cyl, SOHC, 4-cycle	2-cyl, SOHC, 4-cycle	2-cyl, SOHC, 4-cycle	1-cyl, SOHC, 4-cycle	1-cyl, SOHC, 4-cycle	1-cyl, SOHC, 4-cycle	1-cyl, OHV, 4-cycle	1-cyl, OHV, 4-cycle
Capacity:	125cc	125cc	125cc	154cc	49cc	49cc	49cc	49cc	49cc
Power:	7bhp @ 6,000rpm	11.5bhp @ 9,500rpm	15bhp @ 10,500rpm	13.5bhp @ 9,500rpm	4.5bhp @ 9,500rpm	2.3bhp @ 5,700rpm	1.2bhp @ 4,200rpm	1.8bhp @ 5,700rpm	1.8bhp @ 5,700rpm
Transmission:	4-speed manual	4-speed manual	4-speed manual	4-speed manual	3-speed semi-auto	2-speed semi-auto	1-speed	1-speed	1-speed
Frame:	Pressed steel	Pressed steel	Pressed steel	Pressed steel	Pressed steel	Pressed steel	Pressed steel	Pressed steel	Pressed steel
Suspension, Front/Rear:	Earles/S.Arm	Leading-link/S.Arm	Leading-link/S.Arm	Leading-link/S.Arm	Leading-link/S.Arm	Leading-link/S.Arm	Leading-link/S.Arm	Leading-link/Rigid	Telescopic/S.Arm
Brakes, Front/Rear:	Drum/Drum	Drum/Drum	Drum/Drum	Drum/Drum	Drum/Drum	Drum/Drum	Drum/Drum	Drum/Drum	Drum/Drum
Wheels, Front/Rear:	19inch/19inch	16inch/16inch	18inch/18inch	16inch/16inch	17inch/17inch	17inch/17inch	17inch/17inch	19inch/19inch	19inch/19inch
Weight, Dry:	242lbs (110kg)	264lbs (120kg)	242lbs (110kg)	264lbs (120kg)	143lbs (65kg)	119lbs (54kg)	99lbs (45kg)	110lbs (50kg)	110lbs (50kg)
Top Speed:	50mph (80kph)	70mph (113kph)	81mph (130kph)	N/A	43mph (70kph)	31mph (50kph)	19mph (30kph)	31mph (50kph)	31mph (50kph)
Intro Price: (Market'Yr)	N/A	£186-13-4d (UK'62)	£214-2-2d (UK'62)	£196-8-10d (UK'65)	£79gns (UK'62)	N/A	£52-4-9d (UK'67)	£80.00 (UK'69)	N/A

Specifications:	PF50R Amigo	PF50 Novio	PA50 Camino/Hobbit	CZ100 Monkey	Sports Cub C110	S65 Sport 65	S90 Super 90 (CS90)	C200 Touring 90	CT200 Trail 90
Model on display:	1974 PF50R	1975 PF50 Novio	1978 PA50 Hobbit	1965 CZ100 Monkey	1965 Sports Cub C110	1965 S65 Sport 65	1966 S90 Super 90	1964 CA200	1965 CT200 Trail
Year of introduction:	1974	1975	1976	1963	1960	1964	1964	1963	1964
Engine Type:	1-cyl, OHV, 4-cycle	1-cyl, OHV, 4-cycle	1-cyl, 2-cycle	1-cyl, OHV, 4-cycle	1-cyl, OHV, 4-cycle	1-cyl, SOHC, 4-cycle	1-cyl, SOHC, 4-cycle	1-cyl, OHV, 4-cycle	1-cyl, OHV, 4-cycle
Capacity:	49cc	49cc	49cc	49cc	49cc	63cc	89cc	87cc	87cc
Power:	1.8bhp @ 5,700rpm	2bhp @ 6,000rpm	1.8bhp @ 5,700rpm	4.3bhp @ 9,500rpm	5bhp @ 9,500rpm	6.2bhp @ 10,000rpm	8bhp @ 9,500rpm	6.5bhp @ 8,000rpm	6.5bhp @ 8,000rpm
Transmission:	1-speed	1-speed	1-speed	3-speed semi-auto	4-speed manual	4-speed manual	4-speed manual	4-speed manual	4-speed manual
Frame:	Steel tube	Pressed steel	Steel tube	Back bone steel pipe	Pressed steel	Pressed steel	Pressed steel	Pressed steel	Pressed steel
Suspension, Front/Rear:	Telescopic/S.Arm	Telescopic/S.Arm	Telescopic/S.Arm	Rigid/Rigid	Leading-link/S.Arm	Leading-link/S.Arm	Telescopic/S.Arm	Leading-link/S.Arm	Leading-link/S.Arm
Brakes, Front/Rear:	Drum/Drum	Drum/Drum	Drum/Drum	Drum/Drum	Drum/Drum	Drum/Drum	Drum/Drum	Drum/Drum	Drum/Drum
Wheels, Front/Rear:	17inch/17inch	17inch/17inch	17inch/17inch	5inch/5inch	17inch/17inch	17inch/17inch	17inch/17inch	18inch/18inch	17inch/17inch
Weight, Dry:	98lbs (44.5kg)	99lbs (45kg)	99lbs (45kg)	N/A	145lbs (66kg)	170lbs (77kg)	176lbs (80kg)	184lbs (83kg)	181lbs (82kg)
Top Speed:	28mph (45kph)	31mph (50kph)	31mph (50kph)	N/A	53mph (85kph)	60mph (97kph)	62mph (100kph)	N/A	N/A
Intro Price: (Market'Yr)	£119.00 (UK'74)	£119.00 (UK'75)	£159.00 (UK'77)	£67-15-5d (UK'64)	£109-19-0d (UK'62)	£123-2-2d (UK'66)	£142-3-6d (UK'67)	£136-12-2d UK'66)	N/A

Specifications:	CL70 Scrambler	CL90 Scrambler	CM91 Honda 90	CA160 Touring	CB160 Sport 160	CL125A Scrambler	CL160 Scrambler	CL175 Scrambler	CL72 Scrambler 250
Model on display:	1969 CL70-K0	1967 CL90	1969 CM91	1966 CA160 Touring	1965 CB160 Sport	1968 CL125A Scrambler	1966 CL160 Scrambler	1968 CL175-K0	1965 CL72
Year of introduction:	1969	1966	1966	1965	1964	1967	1966	1968	1962
Engine Type:	1-cyl, SOHC, 4-cycle	1-cyl, SOHC, 4-cycle	1-cyl, SOHC, 4-cycle	2-cyl, SOHC, 4-cycle	2-cyl, SOHC, 4-cycle	2-cyl, SOHC, 4-cycle	2-cyl, SOHC, 4-cycle	2-cyl, SOHC, 4-cycle	2-cyl, SOHC, 4-cycle
Capacity:	72cc	89cc	89cc	161cc	161cc	124cc	161cc	174cc	247cc
Power:	5bhp @ 9,000rpm	8bhp @ 9,500rpm	7.5bhp @ 9,500rpm	-	16.5bhp @ 10,000rpm	13.5bhp @ 10,000rpm	16.5bhp @ 10,000rpm	20bhp @ 10,000rpm	24bhp @ 9,000rpm
Transmission:	4-speed manual	4-speed manual	3-speed semi-auto	4-speed manual	4-speed manual	4-speed manual	4-speed manual	5-speed manual	4-speed manual
Frame:	Pressed steel	Pressed steel	Pressed steel	Pressed steel	Steel tube	Steel tube	Steel tube	Steel tube	Steel tube
Suspension, Front/Rear:	Telescopic/S.Arm	Telescopic/S.Arm	Leading-link/S.Arm	Leading-link/S.Arm	Telescopic/S.Arm	Telescopic/S.Arm	Telescopic/S.Arm	Telescopic/S.Arm	Telescopic/S.Arm
Brakes, Front/Rear:	Drum/Drum	Drum/Drum	Drum/Drum	Drum/Drum	Drum/Drum	Drum/Drum	Drum/Drum	Drum/Drum	Drum/Drum
Wheels, Front/Rear:	17inch/17inch	18inch/18inch	17inch/17inch	16inch/16inch	18inch/18inch	18inch/18inch	18inch/18inch	18inch/18inch	19inch/19inch
Weight, Dry:	154lbs (70kg)	202lbs (92kg)	187lbs (85kg)	264lbs (120kg)	N/A	240lbs (109kg)	282lbs (128kg)	262lbs (129kg)	337lbs (153kg)
Top Speed:	N/A	N/A	N/A	N/A	71mph (115kph)	N/A	80mph (119kph)	N/A	N/A
Intro Price: (Market'Yr)	N/A	N/A	N/A	£228-11-8d (UK'65)	N/A	$580.00 (US'66)	N/A	N/A	N/A

Specifications:	SS125 Super Sport	CD/CA175 Touring	CL72 Scrambler 250	Dream CB77	CB77 Super Hawk	CA78 Dream Touring	CB350 Super Sports	Dream CB450-K0	CB450-K1
Model on display:	1969 SS125A	1969 CA175	1965 CL72	1965 CB77 + CYB Kit	1966 CB77	1967 CA78 Touring	1969 CB350-K1	1965 CB450-K0	1968 CB450-K1
Year of introduction:	1967	1969	1962	1961	1961	1963	1968	1965	1968
Engine Type:	2-cyl, SOHC, 4-cycle	2-cyl, SOHC, 4-cycle	2-cyl, SOHC, 4-cycle	2-cyl, SOHC, 4-cycle	2-cyl, SOHC, 4-cycle	2-cyl, SOHC, 4-cycle	2-cyl, SOHC, 4-cycle	2-cyl, DOHC, 4-cycle	2-cyl, DOHC, 4-cycle
Capacity:	124cc	174cc	247cc	305cc	305cc	305cc	325cc	444cc	444cc
Power:	13bhp @ 10,500rpm	17bhp @ 10,500rpm	24bhp @ 9,000rpm	28.5bhp @ 9,000rpm	28.5bhp @ 9,000rpm	23bhp @ 7,500rpm	36bhp @ 10,500rpm	43bhp @ 8,500rpm	43bhp @ 8,500rpm
Transmission:	4-speed manual	4-speed manual	4-speed manual	4-speed manual	4-speed manual	4-speed manual	5-speed manual	4-speed manual	5-speed manual
Frame:	Pressed steel	Steel tube	Steel tube	Steel tube	Steel tube	Pressed steel	Steel tube	Steel tube	Steel tube
Suspension, Front/Rear:	Telescopic/S.Arm	Telescopic/S.Arm	Telescopic/S.Arm	Telescopic/S.Arm	Telescopic/S.Arm	Leading Link/S.Arm	Telescopic/S.Arm	Telescopic/S.Arm	Telescopic/S.Arm
Brakes, Front/Rear:	Drum/Drum	Drum/Drum	Drum/Drum	Drum/Drum	Drum/Drum	Drum/Drum	Drum/Drum	Drum/Drum	Drum/Drum
Wheels, Front/Rear:	16inch/16inch	16inch/16inch	19inch/19inch	18inch/18inch	18inch/18inch	16inch/16inch	18inch/18inch	18inch/18inch	18inch/18inch
Weight, Dry:	216lbs (98kg)	249lbs (113kg)	337lbs (153kg)	350lbs (159kg)	350lbs (159kg)	372lbs (169kg)	328lbs (149kg)	412lbs (187kg)	412lbs (187kg)
Top Speed:	N/A	N/A	N/A	105mph (168kph)	105mph (168kph)	N/A	106mph (170kph)	110mph (180kph)	110mph (180kph)
Intro Price: (Market'Yr)	£189.95 (UK'70)	£220.00 (UK'69)	N/A	£269-19-0d (UK'64)	£269-19-0d (UK'64)	£229-19-0d (UK'64)	N/A	£360-00 (UK'66)	N/A

Specifications:	CB450-K6	CB450D	ST70 Dax	ST90 Trailsport	SS50 Super Sports	CB50J	CB100 Super Sport	CB125S	CT70 Trail
Model on display:	1973 CB450-K6	1968 CB450D	1978 ST70 Dax	1973 ST90-K0	1973 SS50-ZK1	1978 CB50J	1973 CB100-K2	1975 CB125S2	1972 CT70-K1
Year of introduction:	1973	1968	1969	1973	1967	1977	1970	1972	1969
Engine Type:	2-cyl, DOHC, 4-cycle	2-cyl, DOHC, 4-cycle	1-cyl, SOHC, 4-cycle	1-cyl, SOHC, 4-cycle	1-cyl, SOHC, 4-cycle	1-cyl, SOHC, 4-cycle	1-cyl, SOHC, 4-cycle	1-cyl, SOHC, 4-cycle	1-cyl, SOHC, 4-cycle
Capacity:	444cc	444cc	72cc	89cc	49cc	49cc	99cc	122cc	72cc
Power:	43bhp @ 8,500rpm	43bhp @ 8,500rpm	6bhp @ 9,000rpm	6bhp @ 9,000rpm	2.5bhp @ 8,000rpm	2.5bhp @ 7,000rpm	11.5bhp @ 11,000rpm	12bhp @ 9,000rpm	6.2bhp @ 9,000rpm
Transmission:	5-speed manual	4-speed manual	3-speed semi-auto	3-speed semi-auto	4-speed manual	4-speed manual	5-speed manual	5-speed manual	3-speed semi-auto
Frame:	Steel tube	Steel tube	Pressed steel	Pressed steel	Pressed steel	Steel tube	Steel tube	Steel tube	Steel tube
Suspension, Front/Rear:	Telescopic/S.Arm	Telescopic/S.Arm	Telescopic/S.Arm	Telescopic/S.Arm	Telescopic/S.Arm	Telescopic/S.Arm	Telescopic/S.Arm	Telescopic/S.Arm	Telescopic/S.Arm
Brakes, Front/Rear:	Disc/Drum	Drum/Drum	Drum/Drum	Drum/Drum	Drum/Drum	Disc/Drum	Drum/Drum	Drum/Drum	Drum/Drum
Wheels, Front/Rear:	18inch/18inch	18inch/18inch	10inch/10inch	14inch/14inch	17inch/17inch	17inch/17inch	18inch/18inch	18inch/18inch	10inch/10inch
Weight, Dry:	412lbs (187kg)	412lbs (187kg)	143lbs (65kg)	143lbs (65kg)	168lbs (76kg)	167lbs (76kg)	192lbs (87kg)	194lbs (88kg)	143lbs (65kg)
Top Speed:	110mph (180kph)	110mph (180kph)	47mph (75kph)	50mph (80kph)	31mph (50kph)	30mph (48kph)	N/A	N/A	N/A
Intro Price: (Market'Yr)	N/A	N/A	£149.00 (UK'72)	N/A	£122.38 (UK'70)	£319.00 (UK'77)	£200.00 (UK'70)	£199.00 (UK'72)	N/A

Specifications:	CT125 Trail	SL70 Motosport	SL100 Motosport	SL125 Motosport	SL350 Motosport	TL125 Trials	TL250 Trials	TLR200 Reflex	ATC70
Model on display:	1977 CT125'77	1971 SL70-K0	1972 SL100-K2	1972 SL125-K1	1970 SL350-K1	1973 TL125-K0	1976 TL250'76	1986 TLR200'86	1973 ATC70-K0
Year of introduction:	1976	1971	1971	1971	1969	1973	1976	1985	1973
Engine Type:	1-cyl, SOHC, 4-cycle	1-cyl, SOHC, 4-cycle	1-cyl, SOHC, 4-cycle	1-cyl, SOHC, 4-cycle	2-cyl, SOHC, 4-cycle	1-cyl, SOHC, 4-cycle	1-cyl, SOHC, 4-cycle	1-cyl, SOHC, 4-cycle	1-cyl, SOHC, 4-cycle
Capacity:	124cc	72cc	99cc	122cc	325cc	122cc	248cc	195cc	72cc
Power:	9bhp @ 8,000rpm	6.5bhp @ 9,500rpm	11.5bhp @ 11,000rpm	12bhp @ 9,000rpm	30bhp @ 9,500rpm	8bhp @ 8,000rpm	16.5bhp @ 7,000rpm	13.4bhp @ 6,500rpm	-
Transmission:	5-speed manual	4-speed manual	5-speed manual	5-speed manual	5-speed manual	5-speed manual	5-speed manual	6-speed manual	3-speed semi-auto
Frame:	Steel tube	Steel tube	Steel tube	Steel tube	Steel tube	Steel tube	Steel tube	Steel tube	Pressed steel
Suspension, Front/Rear:	Telescopic/S.Arm	Telescopic/S.Arm	Telescopic/S.Arm	Telescopic/S.Arm	Telescopic/S.Arm	Telescopic/S.Arm	Telescopic/S.Arm	Telescopic/S.Arm	Rigid/Rigid
Brakes, Front/Rear:	Drum/Drum	Drum/Drum	Drum/Drum	Drum/Drum	Drum/Drum	Drum/Drum	Drum/Drum	Drum/Drum	Drum/Drum
Wheels, Front/Rear:	19inch/18inch	16inch/14inch	19inch/17inch	21inch/18inch	19inch/18inch	21inch/18inch	21inch/18inch	21inch/18inch	16inch/16inch
Weight, Dry:	249lbs (113kg)	143lbs (65kg)	203lbs (92kg)	201lbs (91kg)	346lbs (157kg)	202lbs (92kg)	218lbs (99kg)	198lbs (90kg)	169lbs (77kg)
Top Speed:	57mph (92kph)	N/A	N/A	63mph (102kph)	84mph (135kph)	N/A	N/A	N/A	N/A
Intro Price: (Market'Yr)	N/A	N/A	N/A	£319.00 (UK'74)	N/A	£329.00 (UK'75)	N/A	£1,135.00 (UK'85)	£395.00 (UK'81)

Specifications:	ATC90 (US90)	ATC250R	CL125S Scrambler	CL175 Scrambler	CL350 Scrambler	CL360 Scrambler	CL450 Scrambler	XL100	XL100S
Model on display:	1972 ATC90-K1	1985 ATC250R'85	1974 CL125-S1	1972 CL125-K6	1972 CL350-K4	1974 CL350-K0	1972 CL450-K5	1974 XL100-K0	1982 XL100S'82
Year of introduction:	1970	1981	1967	1968	1968	1974	1968	1974	1979
Engine Type:	1-cyl, SOHC, 4-cycle	1-cyl, 2-cycle	1-cyl, SOHC, 4-cycle	2-cyl, SOHC, 4-cycle	2-cyl, SOHC, 4-cycle	2-cyl, SOHC, 4-cycle	2-cyl, DOHC, 4-cycle	1-cyl, SOHC, 4-cycle	1-cyl, SOHC, 4-cycle
Capacity:	89cc	246cc	122cc	174cc	325cc	356cc	444cc	99cc	99cc
Power:	-	-	9bhp @ 9,500rpm	20bhp @ 10,000rpm	33bhp @ 9,500rpm	33bhp @ 9,500rpm	43bhp @ 8,000rpm	9.7bhp @ 9,500rpm	10bhp @ 9,500rpm
Transmission:	4-speed semi-auto	5-speed manual	5-speed manual	5-speed manual	5-speed manual	6-speed manual	5-speed manual	5-speed manual	5-speed manual
Frame:	Pressed steel	Steel tube	Steel tube	Steel tube	Steel tube	Steel tube	Steel tube	Steel tube	Steel tube
Suspension, Front/Rear:	Rigid/Rigid	Telescopic/Pro-Link	Telescopic/S.Arm	Telescopic/S.Arm	Telescopic/S.Arm	Telescopic/S.Arm	Telescopic/S.Arm	Telescopic/S.Arm	Telescopic/S.Arm
Brakes, Front/Rear:	Drum/Drum	Disc/Disc	Drum/Drum	Drum/Drum	Drum/Drum	Drum/Drum	Drum/Drum	Drum/Drum	Drum/Drum
Wheels, Front/Rear:	22inch/22inch	23inch/20inch	18inch/17inch	18inch/18inch	19inch/18inch	18inch/18inch	18inch/18inch	19inch/17inch	19inch/16inch
Weight, Dry:	206lbs (93kg)	293lbs (133kg)	196lbs (89kg)	262lbs (119kg)	326lbs (148kg)	326lbs (148kg)	402lbs (182kg)	209lbs (95kg)	176lbs (80kg)
Top Speed:	N/A	N/A	N/A	80mph (129kph)	100mph (161kph)	100mph (161kph)	110mph (177kph)	N/A	N/A
Intro Price: (Market'Yr)	$595.00 (USA'72)	£1,295.00 (UK'82)	N/A	$645.00 (USA'68)	N/A	N/A	N/A	N/A	£574.00 (UK'79)

Specifications:	XL175 Motosport	XL250 Motosport	XL250	XL350	MR50 Elsinore	MR175 Elsinore	MR250 Elsinore	MT125 Elsinore	MT250 Elsinore
Model on display:	1975 XL175-K2	1973 XL250-K0	1975 XL250-K2	1974 XL350-K0	1976 MR50-K1	1976 MR175'76	1976 MR250'76	1974 MT125-K0	1975 MT250-K1
Year of introduction:	1973	1972	1972	1974	1974	1974	1975	1976	1974
Engine Type:	1-cyl, SOHC, 4-cycle	1-cyl, SOHC, 4-cycle	1-cyl, SOHC, 4-cycle	1-cyl, SOHC, 4-cycle	1-cyl, 2-cycle	1-cyl, 2-cycle	1-cyl, 2-cycle	1-cyl, 2-cycle	1-cyl, 2-cycle
Capacity:	173cc	248cc	248cc	348cc	49cc	171cc	248cc	123cc	248cc
Power:	15bhp @ 8,500rpm	20bhp @ 8,000rpm	20bhp @ 8,000rpm	30bhp @ 7,000rpm	-	-	-	13bhp @ 7,000rpm	23bhp @ 6,500rpm
Transmission:	5-speed manual	5-speed manual	5-speed manual	5-speed manual	3-speed manual	5-speed manual	5-speed manual	5-speed manual	5-speed manual
Frame:	Steel tube	Steel tube	Steel tube	Steel tube	Steel tube	Steel tube	Steel tube	Steel tube	Steel tube
Suspension, Front/Rear:	Telescopic/S.Arm	Telescopic/S.Arm	Telescopic/S.Arm	Telescopic/S.Arm	Telescopic/S.Arm	Telescopic/S.Arm	Telescopic/S.Arm	Telescopic/S.Arm	Telescopic/S.Arm
Brakes, Front/Rear:	Drum/Drum	Drum/Drum	Drum/Drum	Drum/Drum	Drum/Drum	Drum/Drum	Drum/Drum	Drum/Drum	Drum/Drum
Wheels, Front/Rear:	21inch/18inch	21inch/18inch	21inch/18inch	21inch/18inch	14inch/12inch	21inch/18inch	21inch/18inch	21inch/18inch	21inch/18inch
Weight, Dry:	239lbs (108kg)	278lbs (126kg)	278lbs (126kg)	321lbs (146kg)	94lbs (42.5kg)	213lbs (97kg)	256lbs (116kg)	211lbs (96kg)	260lbs (118kg)
Top Speed:	N/A	N/A	N/A	N/A	N/A	N/A	N/A	N/A	N/A
Intro Price: (Market'Yr)	N/A	£529.00 (UK'74)	£529.00 (UK'75)	N/A	N/A	N/A	N/A	N/A	N/A

Specifications:	CR250M Elsinore	CB175	CB175 Super Sport	CB200	CB250 Super Sport	CB350 Super Sport	CB250T Dream	CB250N Super Dream	CB400AT Hondamatic
Model on display:	1973 CR250-M0	1969 CB175-K0	1971 CB175-K5	1974 CB200-K0	1968 CB250-K0	1972 CB350-K4	1978 CB250T	1980 CB250N-A	1979 CB400AT
Year of introduction:	1973	1968	1969	1973	1968	1968	1977	1978	1978
Engine Type:	1-cyl, 2-cycle	2-cyl, SOHC, 4-cycle	2-cyl, SOHC, 4-cycle	2-cyl, SOHC, 4-cycle	2-cyl, SOHC, 4-cycle	2-cyl, SOHC, 4-cycle	2-cyl, SOHC, 4-cycle	2-cyl, SOHC, 4-cycle	2-cyl, SOHC, 4-cycle
Capacity:	248cc	174cc	174cc	198cc	249cc	325cc	249cc	249cc	395cc
Power:	33bhp @ 7,500rpm	20bhp @ 10,000rpm	20bhp @ 10,000rpm	17bhp @ 9,000rpm	30bhp @ 10,500rpm	32bhp @ 9,500rpm	27bhp @ 10,000rpm	27bhp @ 10,000rpm	27bhp @ 10,000rpm
Transmission:	5-speed manual	5-speed manual	5-speed manual	5-speed manual	5-speed manual	5-speed manual	5-speed manual	5-speed manual	2-speed Hondamatic
Frame:	Steel tube	Steel tube	Steel tube	Steel tube	Steel tube	Steel tube	Steel tube	Steel tube	Steel tube
Suspension, Front/Rear:	Telescopic/S.Arm	Telescopic/S.Arm	Telescopic/S.Arm	Telescopic/S.Arm	Telescopic/S.Arm	Telescopic/S.Arm	Telescopic/S.Arm	Telescopic/S.Arm	Telescopic/S.Arm
Brakes, Front/Rear:	Drum/Drum	Drum/Drum	Drum/Drum	Drum/Drum	Drum/Drum	Drum/Drum	Disc/Drum	Disc/Drum	Disc/Drum
Wheels, Front/Rear:	21inch/19inch	18inch/18inch	18inch/18inch	18inch/18inch	18inch/18inch	18inch/18inch	18inch/18inch	19inch/18inch	18inch/18inch
Weight, Dry:	213lbs (97kg)	264lbs (120kg)	264lbs (120kg)	291lbs (132kg)	328lbs (149kg)	344lbs (156kg)	364lbs (165kg)	367lbs (166kg)	412lbs (187kg)
Top Speed:	65mph (105kph)	86mph (138kph)	86mph (138kph)	77mph (124kph)	99mph (160kph)	110mph (170kph)	N/A	N/A	N/A
Intro Price: (Market'Yr)	N/A	£280.00 (UK'70)	£269.00 (UK'71)	£429.00 (UK'74)	£289-19-0d (UK'68)	£399.00 (UK'72)	£729.00 (UK'77)	£799.00 (UK'78)	£999.00 (UK'78)

Specifications:	CB500T Twin	Dream CB350F Four	CB400F Super Sport	Dream CB500 Four	CB550 Four K	CB550F Super Sport	Dream CB750 Four	CB750 Four	CB750 Four
Model on display:	1975 CB500T0	1972 CB350F-0	1976 CB400F1	1972 CB500-K1	1978 CB550K'78	1976 CB550F1	1969 CB750	1971 CB750-K1	1972 CB750-K2
Year of introduction:	1975	1972	1975	1971	1974	1975	1969	1969	1969
Engine Type:	2-cyl, DOHC, 4-cycle	4-cyl, SOHC, 4-cycle	4-cyl, SOHC, 4-cycle	4-cyl, SOHC, 4-cycle	4-cyl, SOHC, 4-cycle	4-cyl, SOHC, 4-cycle	4-cyl, SOHC, 4-cycle	4-cyl, SOHC, 4-cycle	4-cyl, SOHC, 4-cycle
Capacity:	498cc	347cc	408cc	498cc	544cc	544cc	736cc	736cc	736cc
Power:	34bhp @ 8,500rpm	34bhp @ 10,000rpm	37bhp @ 8,500rpm	48bhp @ 9,000rpm	50bhp @ 8,500rpm	50bhp @ 8,000rpm	68bhp @ 8,500rpm	68bhp @ 8,500rpm	68bhp @ 8,500rpm
Transmission:	5-speed manual	5-speed manual	6-speed manual	5-speed manual	5-speed manual	5-speed manual	5-speed manual	5-speed manual	5-speed manual
Frame:	Steel tube	Steel tube	Steel tube	Steel tube	Steel tube	Steel tube	Steel tube	Steel tube	Steel tube
Suspension, Front/Rear:	Telescopic/S.Arm	Telescopic/S.Arm	Telescopic/S.Arm	Telescopic/S.Arm	Telescopic/S.Arm	Telescopic/S.Arm	Telescopic/S.Arm	Telescopic/S.Arm	Telescopic/S.Arm
Brakes, Front/Rear:	Disc/Drum	Disc/Drum	Disc/Drum	Disc/Drum	Disc/Drum	Disc/Drum	Disc/Drum	Disc/Drum	Disc/Drum
Wheels, Front/Rear:	19inch/18inch	18inch/18inch	18inch/18inch	19inch/18inch	19inch/18inch	19inch/18inch	19inch/18inch	19inch/18inch	19inch/18inch
Weight, Dry:	425lbs (193kg)	373lbs (169kg)	328lbs (149kg)	443lbs (201kg)	427lbs (193kg)	423lbs (192kg)	481lbs (281kg)	481lbs (281kg)	481lbs (281kg)
Top Speed:	101mph (163kph)	98mph (158kph)	104mph (167kph)	115mph (185kph)	N/A	N/A	125mph (201kph)	125mph (201kph)	125mph (201kph)
Intro Price: (Market'Yr)	£699.00 (UK'75)	N/A	£669.00 (UK'75)	£629.00 (UK'72)	£995.00 (UK'77)	£879.00 (UK'76)	£695.00 (UK'69)	£719.00 (UK'71)	£761.00 (UK'72)

Specifications:	CB750 Four	CB750 Four	CB750A Hondamatic	CB750K LTD Edition	CB750F Super Sport	CB750F Super Sport	CB750F Phil Read Rep.	GL1000 Gold Wing	GL1000 G.Wing Ltd
Model on display:	1976 CB750'76 (K6)	1978 CB750-K8	1976 CB750A'76	1979 CB750K'79	1976 CB750-F1	1977 CB750F'77 (F2)	1978 CB750-F2	1975 GL1000-K0	1976 GL1000-LTD'76
Year of introduction:	1969	1969	1976	1979	1975	1977	1978	1974	1976
Engine Type:	4-cyl, SOHC, 4-cycle	4-cyl, SOHC, 4-cycle	4-cyl, SOHC, 4-cycle	4-cyl, DOHC, 4-cycle	4-cyl, SOHC, 4-cycle	4-cyl, SOHC, 4-cycle	4-cyl, SOHC, 4-cycle	4-cyl, SOHC, 4-cycle	4-cyl, SOHC, 4-cycle
Capacity:	736cc	736cc	736cc	749cc	736cc	736cc	736cc	999cc	999cc
Power:	68bhp @ 8,500rpm	68bhp @ 8,500rpm	47bhp @ 7,500rpm	77bhp @ 9,000rpm	67bhp @ 8,500rpm	73bhp @ 9,000rpm	73bhp @ 9,000rpm	80bhp @ 7,500rpm	80bhp @ 7,500rpm
Transmission:	5-speed manual	5-speed manual	2-speed Hondamatic	5-speed manual	5-speed manual	5-speed manual	5-speed manual	5-speed manual	5-speed manual
Frame:	Steel tube	Steel tube	Steel tube	Steel tube	Steel tube	Steel tube	Steel tube	Steel tube	Steel tube
Suspension, Front/Rear:	Telescopic/S.Arm	Telescopic/S.Arm	Telescopic/S.Arm	Telescopic/S.Arm	Telescopic/S.Arm	Telescopic/S.Arm	Telescopic/S.Arm	Telescopic/S.Arm	Telescopic/S.Arm
Brakes, Front/Rear:	Disc/Drum	Disc/Drum	Disc/Drum	Disc/Drum	Disc/Disc	Disc/Disc	Disc/Disc	Disc/Disc	Disc/Disc
Wheels, Front/Rear:	19inch/18inch	19inch/18inch	19inch/18inch	19inch/18inch	19inch/18inch	19inch/18inch	19inch/18inch	19inch/17inch	19inch/17inch
Weight, Dry:	481lbs (281kg)	508lbs (231kg)	578lbs (262kg)	508lbs (231kg)	538lbs (244kg)	538lbs (244kg)	N/A	584lbs (265kg)	584lbs (265kg)
Top Speed:	125mph (201kph)	125mph (201kph)	97mph (156kg)	124mph (200kph)	125mph (201kph)	125mph (201kph)	125mph (201kph)	121mph (193kph)	121mph (193kph)
Intro Price: (Market'Yr)	£1,145.00 (UK'76)	N/A	N/A	N/A	£1,079.00 (UK'75)	£1,379.00 (UK'77)	£1,895.00 (UK'78)	£1,600.00 (UK'76)	N/A

Specifications:	GL1000 G.Wing Exec	GL1000 Gold Wing	GL1100 Gold Wing DX	GL1200 G.Wing Ltd	CBX	CBX	CBX	Z50R Christmas	MB5/MB50
Model on display:	1977 GL1000-K1	1979 GL1000K-Z	1982 GL1100D-C	1985 GL1200L'85	1979 CBX-Z	1980 CBX'80 (CBX-A)	1981 CBX'81 (CBX-B)	1986 Z50RD'86 Special	1982 MB50'82
Year of introduction:	1977	1979	1980	1985	1978	1980	1981	1986	1979
Engine Type:	4-cyl, SOHC, 4-cycle	4-cyl, SOHC, 4-cycle	4-cyl, SOHC, 4-cycle	4-cyl, SOHC, 4-cycle	6-cyl, DOHC, 4-cycle	6-cyl, DOHC, 4-cycle	6-cyl, DOHC, 4-cycle	1-cyl, SOHC, 4-cycle	1-cyl, 2-cycle
Capacity:	999cc	999cc	1085cc	1182cc	1047cc	1047cc	1047cc	49cc	49cc
Power:	80bhp @ 7,500rpm	78bhp @ 7,000rpm	81bhp @ 7,500rpm	94bhp @ 7,000rpm	105bhp @ 9,000rpm	100bhp @ 9,000rpm	100bhp @ 9,000rpm		7bhp @ 9,000rpm
Transmission:	5-speed manual	5-speed manual	5-speed manual	5-speed manual	5-speed manual	5-speed manual	5-speed manual	3-speed semi-auto	5-speed manual
Frame:	Steel tube	Steel tube	Steel tube	Steel tube	Steel tube	Steel tube	Steel tube	Steel tube	Steel tube
Suspension, Front/Rear:	Telescopic/S.Arm	Telescopic/S.Arm	Telescopic/S.Arm	Telescopic/S.Arm	Telescopic/S.Arm	Telescopic/S.Arm	Telescopic/S.Arm	Telescopic/S.Arm	Telescopic/S.Arm
Brakes, Front/Rear:	Disc/Disc	Disc/Disc	Disc/Disc	Disc/Disc	Disc/Disc	Disc/Disc	Disc/Disc	Drum/Drum	Disc/Drum
Wheels, Front/Rear:	19inch/17inch	19inch/17inch	19inch/17inch	16inch/15inch	19inch/18inch	19inch/18inch	19inch/18inch	8inch/8inch	18inch/18inch
Weight, Dry:	N/A	N/A	679lbs (308kg)	782lbs (355kg)	545lbs (247kg)	555lbs (252kg)	600lbs (272kg)	109lbs (49kg)	174lbs (79kg)
Top Speed:	121mph (193kph)	N/A	N/A	100mph (243kph)	134mph (216kph)	130mph (209kph)	130mph (209kph)	N/A	30mph (48kph)
Intro Price: (Market'Yr)	£2,300.00 (UK'77)	£2,399.00 (UK'79)	£3,088.00 (UK'81)	N/A	£2,560.00 (UK'78)	N/A	£3,395.00 (UK'82)	N/A	£399.00 (UK'80)

Specifications:	C70	CG125	CD200T Benly	CX500	CX650 Turbo	FT500 Ascot	XBR500	NT650 Hawk GT	VTR250F Interceptor
Model on display:	1982 C70'82 Passport	1980 CG125-K1	1981 CD200T-B	1978 CX500'78	1983 CX650TC'83	1982 FT500'82	1986 XBR500-G	1988 NT650'88	1989 VTR250F'89
Year of introduction:	1969	1975	1979	1977	1983	1982	1985	1988	1988
Engine Type:	1-cyl, SOHC, 4-cycle	1-cyl, OHV, 4-cycle	2-cyl, SOHC, 4-cycle	2-cyl, OHV, 4-cycle	2-cyl, OHV, 4-cycle	1-cyl, SOHC, 4-cycle	1-cyl, SOHC, 4-cycle	2-cyl, SOHC, 4-cycle	2-cyl, DOHC, 4-cycle
Capacity:	72cc	124cc	194cc	496cc	673cc	498cc	498cc	647cc	249cc
Power:	7bhp @ 9,000rpm	11bhp @ 9,000rpm	15bhp @ 8,500rpm	50bhp @ 9,000rpm	100bhp @ 8,000rpm	35bhp @ 6,500rpm	44bhp @ 7,000rpm	58bhp @ 5,000rpm	45bhp @ 13,500rpm
Transmission:	3-speed semi-auto	4-speed manual	4-speed manual	5-speed manual	5-speed manual	5-speed manual	5-speed manual	5-speed manual	6-speed manual
Frame:	Pressed steel	Steel tube	Steel tube	Steel tube	Steel tube	Steel tube	Steel tube	Aluminium box	Steel tube
Suspension, Front/Rear:	Leading-link/S.Arm	Telescopic/S.Arm	Telescopic/S.Arm	Telescopic/S.Arm	Telescopic/S.Arm	Telescopic/S.Arm	Telescopic/S.Arm	Telescopic/Pro-Arm	Telescopic/Pro-Arm
Brakes, Front/Rear:	Drum/Drum	Drum/Drum	Drum/Drum	Disc/Drum	Disc/Disc	Disc/Disc	Disc/Drum	Disc/Disc	Disc/Disc
Wheels, Front/Rear:	17inch/17inch	18inch/18inch	17inch/17inch	19inch/18inch	18inch/17inch	18inch/18inch	19inch/18inch	18inch/18inch	18inch/18inch
Weight, Dry:	158lbs (72kg)	251lbs (114kg)	269lbs (122kg)	441lbs (200kg)	518lbs (235kg)	351lbs (159kg)	331lbs (150kg)	393lbs (178kg)	331lbs (150kg)
Top Speed:	35mph (56kph)	N/A	N/A	106mph (171kph)	140mph (225kph)	N/A	N/A	N/A	N/A
Intro Price: (Market'Yr)	N/A	£349.00 (UK'77)	£751.00 (UK'81)	£1,249.00 (UK'78)	£3,450.00 (UK'83)	£1,350.00 (UK'82)	£1,749.00 (UK'85)	N/A	N/A

Specifications:	CB700SC Nighthawk S	CB750F Super Sport	CB900F Super Sport	CB1100F Super Sport	CB1100R	VF500F Interceptor	VF750F Interceptor	VF1000R	VFR750F
Model on display:	1984 CB700SC'84	1980 CB750F'80	1982 CB900F'82	1983 CB1100F'83	1981 CB1100R-B	1986 VF500F'86	1983 VF750F'83	1985 VF1000R'85	1990 VFR750F'90
Year of introduction:	1984	1979	1979	1983	1981	1984	1983	1984	1986
Engine Type:	4-cyl, DOHC, 4-cycle	4-cyl, DOHC, 4-cycle	4-cyl, DOHC, 4-cycle	4-cyl, DOHC, 4-cycle	4-cyl, DOHC, 4-cycle	V4-cyl, DOHC, 4-cycle	V4-cyl, DOHC, 4-cycle	V4-cyl, DOHC, 4-cycle	V4-cyl, DOHC, 4-cycle
Capacity:	696cc	748cc	901cc	1062cc	1062cc	498cc	748cc	998cc	748cc
Power:	80bhp @ 10,000rpm	78bhp @ 9,000rpm	91bhp @ 9,000rpm	108bhp @ 9,000rpm	120bhp @ 9,000rpm	68bhp @ 11,500rpm	86bhp @ 9,500rpm	125bhp @ 10,000rpm	102bhp @ 9,500rpm
Transmission:	6-speed manual	5-speed manual	5-speed manual	5-speed manual	5-speed manual	6-speed manual	5-speed manual	5-speed manual	6-speed manual
Frame:	Steel tube	Steel tube	Steel tube	Steel tube	Steel tube	Steel box-section	Steel box-section	Steel box-section	Aluminium twin-spar
Suspension, Front/Rear:	Telescopic/S.Arm	Telescopic/S.Arm	Telescopic/S.Arm	Telescopic/S.Arm	Telescopic/S.Arm	Telescopic/Pro-Link	Telescopic/Pro-Link	Telescopic/Pro-Link	Telescopic/Pro-Link
Brakes, Front/Rear:	Disc/Drum	Disc/Disc	Disc/Disc	Disc/Disc	Disc/Disc	Disc/Disc	Disc/Disc	Disc/Disc	Disc/Disc
Wheels, Front/Rear:	16inch/16inch	18inch/18inch	18inch/18inch	18inch/17inch	18inch/18inch	16inch/18inch	16inch/18inch	16inch/18inch	17inch/17inch
Weight, Dry:	474lbs (215kg)	503lbs (228kg)	534lbs (242kg)	534lbs (242kg)	514lbs (233kg)	406lbs (184kg)	481lbs (281kg)	549lbs (249kg)	476lbs (216kg)
Top Speed:	N/A	N/A	N/A	N/A	143mph (230kph)	N/A	138mph (222kph)	150mph (241kg)	N/A
Intro Price: (Market'Yr)	$3,398.00 (USA'84)	£1,780.00 (UK'80)	£2,099.00 (UK'79)	N/A	N/A	£2,550.00 (UK'84. F2)	£2,575.00 (UK'83)	£5,250.00 (UK'84)	£3,649.00 (UK'86)

Specifications:	CBR600F Hurricane	CBR1000F Hurricane	CBR900RR Fireblade
Model on display:	1989 CBR600F'89	1987 CBR1000F'87	1992 CBR900RR-N
Year of introduction:	1987	1987	1992
Engine Type:	4-cyl, DOHC, 4-cycle	4-cyl, DOHC, 4-cycle	4-cyl, DOHC, 4-cycle
Capacity:	598cc	998cc	893cc
Power:	85bhp @ 11,000rpm	135bhp @ 9,250rpm	122bhp @ 10,500rpm
Transmission:	6-speed manual	6-speed manual	6-speed manual
Frame:	Steel box-section	Steel box-section	Aluminium twin-spar
Suspension, Front/Rear:	Telescopic/Pro-Link	Telescopic/Pro-Link	Telescopic/Pro-Link
Brakes, Front/Rear:	Disc/Disc	Disc/Disc	Disc/Disc
Wheels, Front/Rear:	17inch/17inch	17inch/17inch	16inch/17inch
Weight, Dry:	397lbs (180kg)	549lbs (249kg)	408lbs (185kg)
Top Speed:	142mph (229kph)	154mph (248kg)	164mph (264kph)
Intro Price: (Market'Yr)	£3,299.00 (UK'87)	£4,399.00 (UK'87)	N/A

THE DAVID SILVER HONDA COLLECTION
150 models from 1950s to 1990s

Visiting Us

Thank you for acquiring a copy of this guide book, we hope you found the content interesting and informative.

If you have already visited our museum then as you strolled through both floors of the collection I do hope you enjoyed the 150 plus classic Honda motorcycles on display and that the timeline story in this book and on the walls of the museum helped provide a background to the amazing early decades of the world famous Honda marque.

The timeline story sets out to cover the company's global activity as seen from Japan with focus primarily on the motorcycle range of products. Some models referenced may not have been destined for the UK market as hundreds of models in different specifications and naming to suit different market requirements were produced over time. Some models although almost identical had a different name/code in one market to another. Also, the model launch date stated may be a year earlier than the actual on-sale date, allowing some model years to be easily confused!

Like any historical presentation we of course may have got something wrong! Do let us know if you think any fact is possibly incorrect, we do however hope you found our Honda timeline of great interest.

Our Honda parts shop shares the same opening times as the museum so is always available to help with your parts requirements. It is also always open online if you require any parts now or in the future.

If you have yet to visit our museum or simply would like to return for a second visit then all the required details are set out overleaf.

Thank you for your interest and happy Honda riding.

David Silver

Opening times:
Monday to Friday from 09:00 to 17:30, Saturday from 09:00 to 16:00.
Last admission is one hour before closing.
Closed on Sundays and all UK bank holidays, from 21st Dec to 1st Jan and Easter, incl. Easter Saturday.

Groups welcome by prior arrangement.

How to find us:
Our Honda Collection and parts warehouse/counter are located together in Leiston, a town in eastern Suffolk, England.
It is situated near Saxmundham and Aldeburgh, about 2 miles from the North Sea coast and is 21 miles northeast of Ipswich and 90 miles northeast from London.

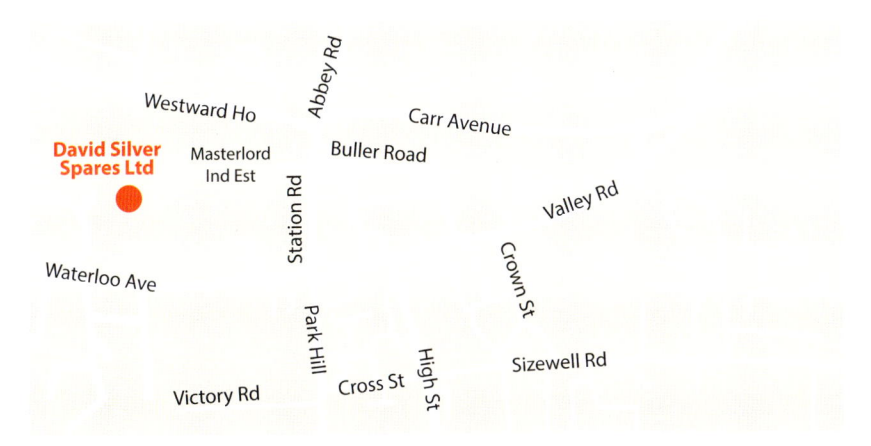

Facilities:
The following are available during your visit:
- Car parking.
- Toilets (including disabled).
- Disabled access and chair lift to the first floor.
- Classic Japanese bikes for sale.
- Hot drinks machine.

Contacts:
Phone: +44(0)1728 833 020
Website: www.davidsilverhondacollection.co.uk
Facebook: www.facebook.com/SilverSpares
Twitter: twitter.com/D_Silver_Spares

Address: The David Silver Honda Collection
           Unit 14, Masterlord Industrial Estate
           Station Road, Leiston, Suffolk
           IP16 4JD, England.

We look forward to welcoming you to our Honda Collection, showroom and parts counter.